Solar Power Generation: Technology, New Concepts & Policy

Solar Power Generation: Technology, New Concepts & Policy

Editor

Prashant Salunke

scitus academics

Solar Power Generation: Technology, New Concepts & Policy

Edited by **Prashant Salunke**

Printed in 2017

ISBN: 978-1-68117-039-8

Library of Congress Control Number: 2015931840

SCITUS Academics LLC,
616, Corporate Way, Suite 2, 4766,
Valley Cottage, NY 10989

www.scitusacademics.com

Contents

Preface

Energy resources consist of fossil fuel such as oil, coal, and natural gas, and non-fossil energy such as solar, hydraulic, and geothermal energy. The supply of fossil fuels, the primary energy source today, is finite, with concerns of depletion, while its combusted gas is destructive to the global environment. These problems have prompted the recent focus on non-fossil energy.

These non-fossil energies do not release environment-polluting substances such as CO2, and because of their ability to be used sustainably, they are called recyclable energy. The most anticipated of them all is solar energy. It is said that if we could convert 100% of one second of sunlight that reaches Earth into solar energy, we could produce all the energy the entire world needs in one hour. That is how immense solar energy is, and considering that it is also clean, it is an energy source with great potential.

Solar power is produced by collecting sunlight and converting it into electricity. This is done by using solar panels, which are large flat panels made up of many individual solar cells. It is most often used in remote locations, although it is becoming more popular in urban areas as well.

This book offers a global perspective of the current state of affairs in the field of solar power engineering.

Editor

Chapter 1

Metal Oxide Nanomaterials, Conducting Polymers and Their Nanocomposites for Solar Energy

Sadia Ameen[1], M. Shaheer Akhtar[2], Minwu Song[1] and Hyung Shik Shin[1]

[1]Energy Materials & Surface Science Laboratory, Solar Energy Research Center, School of Chemical Engineering, Chonbuk National University, Jeonju, Republic of Korea

[2]New and Renewable Energy Material Development Center (NewREC), Chonbuk National University, Jeonbuk, Republic of Korea

INTRODUCTION

The increasing concern on energy and the global warming due to the depletion of fossil fuel demands to search the alternative renewable energy resources for covering the energy crisis in the coming decade.

A very popular renewable source called photovoltaic device is anticipated to solve energy problem, which converts directly the solar energy from sun to the electricity energy. Recently, dye sensitized solar cells (DSSCs) are widely used as promising photovoltaic device owing to its important properties like high solar to electricity energy conversion efficiency, low production cost, ease of fabrication and vast varieties of various semiconducting materials. DSSC is composed of few micrometer-thick nanocrystalline semiconducting oxides thin film combined with monolayer of charge-transfer dye as a photoanode, a redox electrolyte and a platinum metal electrode as counter electrode. In principle, upon illumination, the electron injection to conduction band of semiconductor takes place by the absorption of photons from dye molecules and the redox electrolyte regenerates the oxidized dye by the transportation of charges or ions. These days, the photovoltaic devices are facing inherent drawbacks such as leakage and evaporation problem that limits its practical application. In this regards, efforts are being done to overcome the leakage and evaporation of liquid electrolyte with solid or gel electrolytes such as room temperature molten salts (RTMSs), p-type semiconductor, conducting organic polymers and polymer gel electrolytes. Furthermore, the choice of catalytic in counter component of DSSCs is crucial to improve the reduction rate of I_3^- to I^- in the redox electrolyte. In general, the conducting glass electrode without any catalytic materials such as metals, conducting polymers etc shows very low electrocatalytic activity towards the iodide couple electrolyte due to overvoltage and high energy loss. It has been realized that the low resistance and high electrocatalytic materials might deliver the better catalyst to avoid the overvoltage and energy loss for the high reduction of I_3^- in redox electrolyte.

The semiconducting nanomaterials based thin film electrodes with high surface area are of great significance for acquiring the high amount dye adsorption which leads to the higher light harvesting efficiency and photocurrent density. Various metal oxides semiconductors such as TiO_2, ZnO and SnO_2 have shown good optical and electronic properties and are accepted as the effective photoanode materials for DSSCs. Additionally, the morphology and sizes of metal oxides semiconducting materials, particularly one dimensional (1D) nanostructures like nanorods (NRs), nanowires (NWs) and nanotubes (NTs) based electrodes have shown increased electron transfer during the operation of DSSCs owing to their high surface to volume ratio and arrangements. Until now, the photoanodes with TiO_2, SnO_2/ZnO, Nb_2O_5 and ZnO nanomaterials have presented the maximum solar to electricity energy conversion efficiencies of ~11.2%, ~8%, ~2% and ~5% respectively.

On the other hand, the conducting polymers are regarded as promising semiconducting materials due to distinguishable electrical properties, mechanical flexibility, and the relative ease of processing. The conductive polymers such as polypyrrole, poly (3, 4-ethylenedioxythiophene) and polyaniline (PANI) are frequently used in DSSC as hole transport materials, electron acceptor and electrocatalytic materials for tri-iodide reduction in redox electrolyte. Among them, PANI is an excellent host for the trapping of semiconducting nanomaterials and conducts the electric charges through the polymeric chain due to extended ϖ-electron conjugation. The conducting polymers and dye sensitized metal oxides are good electron donors upon the photo-excitation during the operation of DSSCs.

In this chapter, we have briefly discussed the different conducting polymers, metal oxides and their application for the performance of DSSCs. The chapter includes the brief literature surveys, properties and photovoltaic properties of various metal oxides nanomaterials, nanofillers in polymer electrolytes and the conducting polymers. Additionally, the latest research advancements are surveyed for the development of efficient conducting polymers as p-type semiconducting nanomaterials for counter electrode materials and efficient nanofillers in the solid polymers of DSSCs. Moreover, the doping and the utilization of TiO_2 and ZnO nanomaterials for the performance of DSSCs have been discussed in details. It has been seen that the preparation methods, doping, morphologies, and the sizes of conducting polymers and metal oxides have shown the considerable impact on the electrical properties of the nanomaterials and performances of DSSCs. The study also demonstrates the enhanced properties of inorganic metal oxides like ZnO and TiO_2 with different sizes and morphologies for achieving the efficient photovoltaic properties of DSSCs such as J_{SC}, V_{OC}, FF and conversion efficiency.

TYPES OF CONDUCTING POLYMERS

The conducting polymers are composed of ϖ-conjugated polymeric chain and are known as "synthetic metals" [1-2]. These extended ϖ-conjugated systems of conducting polymers have alternating single and double bonds along the polymeric chain [3]. The conducting polymers display the overlapping of molecular orbital to allow the formation of delocalized molecular wave functions and secondly these molecular orbital must be partially filled so that there is a free movement of electrons in the polymeric structure. The presence of unusual electronic properties such as electrical conductivity, low ionization potential and high electron affinity are associated with the ϖ-electron backbone of the conjugated polymers. These are promising candidates for electronics applications, and offer possible replacements of the conventional metals and inorganic semiconductors [4-5]. The electrical conductivity of the conducting polymers could be altered upon partial oxidation or reduction by a commonly referred process called 'doping'. The electrical conductivity of these polymers could be changed from insulating to metallic by chemical or electrochemical doping and they could be used to produce electronic devices. These polymers have the electrical properties like that of metals, and have attractive characteristics of organic polymers such as light weight, resistance to corrosion, flexibility and lower cost. Additionally, these polymers could be tailor-made to the requirements of the application through modifications in the polymer structure by varying the functional groups in the organic moiety. The commercial applications of conducting polymers are in thin film transistor, batteries, antistatic coatings, electromagnetic shielding, artificial muscles, light-emitting diodes, gas and bio-sensors [6], fuel and solar cells, fillers [7] and corrosion protective coatings [8]. The conducting polymers are easy to synthesize through chemical or electrochemical processes, and their molecular chain structure could be modified conveniently by the copolymerization or structural derivations. Typically, the conducting polymers are of several types as listed below:

Polypyrrole (PPY)

Polypyrrole (PPy) is a versatile polymer of significant properties like redox activity [9] ion-exchange, ion discrimination capacities [10], electrochromic effects, charge/discharge processes [11] and exhibits strong absorptive properties towards gases [12], catalytic activity [13] and corrosion protection properties [14]. It is one of the important conducting polymers due to its good electrochemical reversibility between its conducting and insulating states and the ease of preparation through chemical or electrochemical routes [15].

Poly Phenylenes) (PP)

Poly phenylene (PP) is one of the most unusual electro conducting polymers due to the extended planar conjugated ϖ-system, along with high strength and high heat resistance [16]. The most widely used method of PP production is benzene oxidation with a Friedel-Crafts catalyst (the Kovacic method) [17], which yield a polycrystalline powder. Besides, electrochemical polymerization is also a method for PP synthesis, but the molecular weight of the polymer is limited due to its insolubility and chemical defect [18].

POLYACETYLENE (PA)

Polyacetylene (PA) is the polymer of highest conductivity as compared to those of conventional metals. PA has the simplest structure of the linear chains of C-H units with alternating single and double bonds [19]. Moreover, the existence of the two hydrogen atoms in its repeat unit offers ample opportunity to decorate the backbone with pendants which perturbs the electronic conjugation and influences the molecular alignment of the polymeric chain. Significantly, the proper structural design might tune the backbone-pendant interplay into harmony and synergy, generating new substituted PAs with novel functionalities [20].

Polyazule (PAZ)

The electron-donor and electron-acceptor character of polyazulene (PAz) has been explained by the electron-donor effect of the seven-membered ring toward the five-membered ring. The five-membered ring carries a partial negative charge and the seven-membered ring of azulene carries a partial positive charge. The polymers and its derivatives show high electrical conductivity almost similar as polythiophene, polypyrrole and polyaniline [21].

Polyindole (PIN)

Polyindol (PIN) is an electroactive polymer which could be obtained by the anodic oxidation of indole in various electrolytes. It is reported that the conductivity of PIN is lower than that of PPy and PANI but its thermal stability is better with respect to PANI and PPy. PIN, a macromolecular compound, is a good candidate for applications in various areas, such as electronics, electrocatalysis, and active materials for anodes of batteries, anticorrosion coatings and pharmacology.

Polycarbazole (PCZ)

Polycarbazole (PCz) among conducting polymers, is attributed with good electroactivity, and useful thermal, electrical and photophysical properties [22]. However, ϖ-ϖ electron system along its backbone imparts rigidity to the polymer and therefore, makes it infusible and poorly processable. The increasing interest in PCz is towards its role as a hole-transport material and an efficient photoluminescence unit [23]. Derivatives of carbazole are easily prepared by the substitution at -N atom and thus, the solubility and functionality of the resulting polymers could be improved. More importantly, the substituted groups might influence the effective conjugation length which is promising materials in making the emitting light in devices.

Polyaniline (PANI)

Polyaniline (PANI) exhibits the high stability, conductivity and low cost [24-25]. PANI basically undergoes oxidative polymerization in the presence of a protonic acid. Protonation induces an insulator-to-conductor transition, while the number of ϖ-electrons in the chain remains constant. The oxidation and reduction takes place on this –NH– group, and various forms are obtained due to the number of imine and amine segments on the PANI chain. Other substituted aniline like *N*-benzenesulfaniline [26], o-, p- and m-toluidine, o-chloroaniline [27], o-, m- and p-halogenated anilines [28] and 1-Napthylamine are also the subject of current studies and could be used for the semiconducting polymers based electronic applications.

Out of several conducting polymers, the interest of researchers in PANI could possibly be linked to the numerous applications that exist for the electronic conducting polymers and also aniline is cheap product and also a very stable material. On the other hand, the nanocomposites of conducting polymers with inorganic semiconducting nanomaterials show the improved mechanical, electrical and thermal properties due to the combined effects of both the semiconducting nanomaterials and conducting polymers. In particular, PANI nanocomposites display applications on a large scale for various electrochemical, electrorheological and in the electronic fields such as batteries, sensors, controlling systems and organic displays [29]. The nanocomposites of PANI with cadmium sulphide has been discussed in the next section of the chapter

NANOCOMPOSITES OF CONDUCTING POLYMERS

Nanocomposites of Pani and Cadmium Sulphide

Cadmium Sulphide (CdS) is a semiconductor with a direct band gap of ~2.42 eV which displays superior optical, photophysical and photochemical properties [30]. The nanocomposites of CdS and PANI have presented the effective electrode materials for many electrochemical, photoelectrochemical, sensing and electrochromic devices [31]. The nanocomposites are anticipated as effective and promising electrode materials in many electrochemical devices. Xi et al studied the influence of optical and absorption properties of CdS by the incorporation of CdS into PANI matrix [32]. R. Seoudi et al studied the dependence of structural, vibrational spectroscopy and optical properties on the particle sizes of PANI/CdS nanocomposites [33]. B.T. Raut et al reported the novel method of fabrication of PANI/CdS nanocomposites and studied the structural, morphological and optoelectronic properties [34]. In this regards, Ameen et al has reported a simple solution method to synthesize the CdS decorated PANI nanorods (NRs) and studied the electrochemical impedance properties of the nanocomposites [35].

The synthesized PANI NRs exhibit the entangled network with diameter of ~40-50 nm and length of several hundred nanometers, as shown in Fig. 1 (a). The uniform decoration and the thicknesses of CdS-PANI NRs increase gradually with the increase of $CdCl_2$ concentration. After sensitization with the highest concentration of $CdCl_2$ (0.1 M), the surface of PANI NRs (Fig. 1(b)) is completely decorated by CdS nanoparticles which results in the enhanced diameter of ~60-70 nm. The TEM characterization (Fig. 1 (c and d)) clearly justifies the decoration of CdS nanomaterials on the surface of PANI NRs and shows the increased thickness of PANI NRs with the average diameter of ~60-70 nm due to the decoration of CdS nanoparticles with highest concentration of $CdCl_2$ (0.1 M). From the EDS studies, the overall CdS contents have been estimated as 0.34, 0.53 and 1.08 wt% in the synthesized CdS-PANI NRs with $CdCl_2$ concentrations of 0.01 M, 0.05 M and 0.1 M respectively.

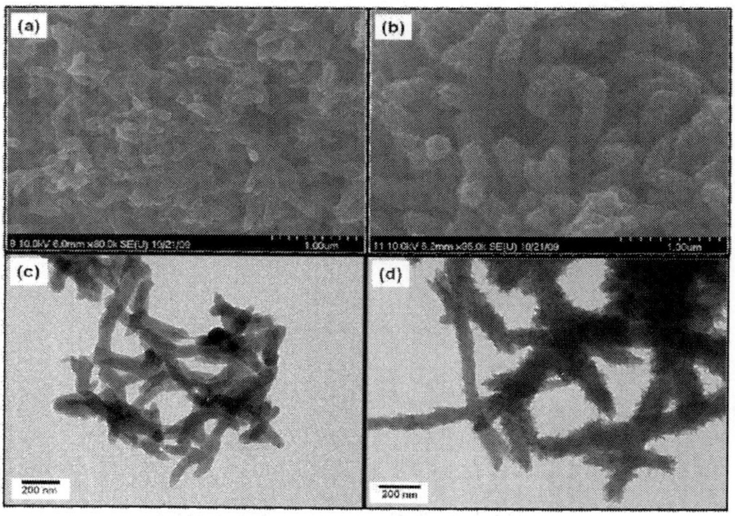

Figure 1: FESEM (a, b) and TEM (c, d) images of synthesized PANI NRs and CdS-PANI NRs. Reprinted with permission from [Ameen S., 2012], Chem. Eng. J. 181 (2012) 806 ©2012, Elsevier Ltd.

The Raman bands at ~1175 cm^{-1}, ~1507 cm^{-1}, ~1595 cm^{-1} are observed in all the samples of CdS-PANI NRs (Fig. 2), corresponding to the C-H bending vibration of the semi quinonoid rings (cation-radical segments), N-H deformation vibration associated with the semiquinonoid structures and C=C stretching vibration in the quinonoid ring respectively [36]. PANI NRs show a relatively high band at ~1368 cm^{-1} in the spectrum which provides the information of the $C\text{-}N^{+\bullet}$ vibration of delocalized polaronic structures [37]. The absence of this band in CdS-PANI NRs might due to the efficient interaction of imine (-NH) group of PANI with CdS nanomaterials.

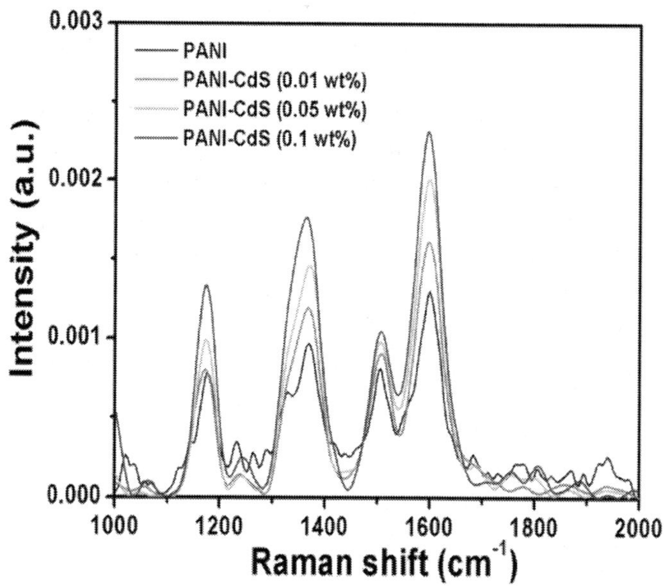

Figure 2: Raman Spectra of PANI NRs and CdS-PANI NRs. Reprinted with permission from [Ameen S., 2012], Chem. Eng. J.181 (2012) 806 ©2012, Elsevier Ltd.

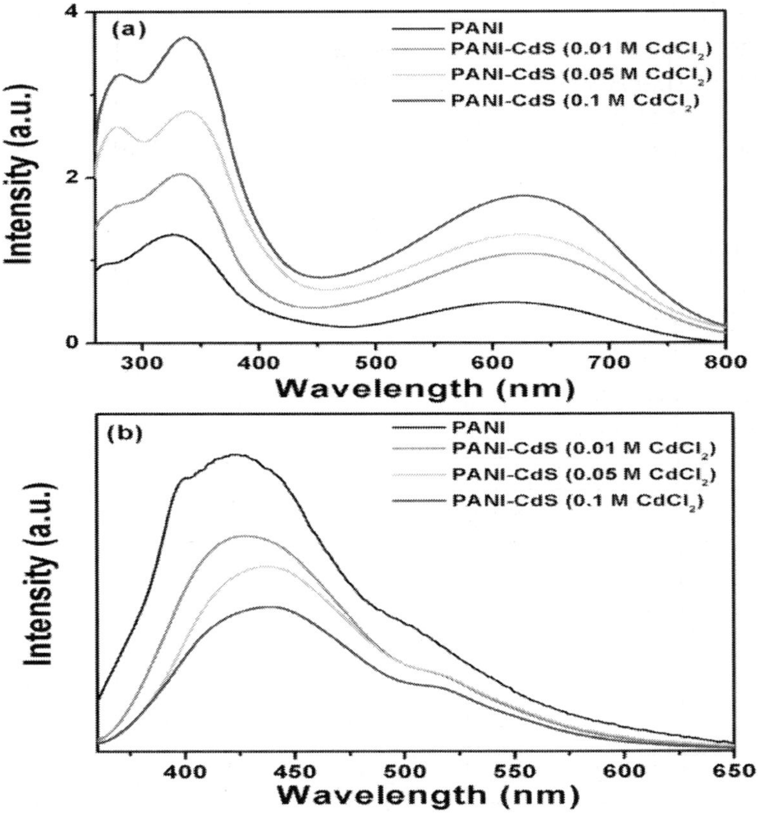

Figure 3: UV-vis absorption spectra (a) and photoluminescence spectra (b) of PANI NRs and CdS-PANI NRs. Reprinted with permission from [Ameen S., 2012], Chem. Eng. J.181 (2012) 806 ©2012, Elsevier Ltd.

The UV-Vis absorption spectra of the PANI NRs and CdS-PANI NRs are depicted in Fig. 3(a). In PANI NRs, the broad absorption bands at ~617 nm is related to n-ϖ* transition, and the absorption peak at ~268 nm and ~327 nm arises due to ϖ-ϖ* transition within the benzenoid segment which is associated to the extent of conjugation between adjacent phenyl rings in the PANI [38]. On comparison to PANI NRs, the red shifting are seen and absorption bands move to higher wavelength of ~279 nm, ~338 nm and ~630 nm respectively in CdS-PANI NRs due to the sensitization of CdS nanomaterials with PANI NRs. The red shift of absorption bands with high intensities reveals that PANI NRs might form a partial bond with CdS nanoparticles. The room temperature photoluminescence (PL) spectra (Fig. 3(b)) of PANI NRs and CdS-PANI NRs exhibit a single large amplitude band in the blue green region which originated due to the ϖ–ϖ* transition of the benzenoid unit of PANI [39]. The sensitization of PANI NRs with the highest concentration of $CdCl_2$ (0.1 M) causes a significant red shift from ~421 nm to ~438 nm as compared to the PANI NRs which might occur by the chemical interaction between -NH groups of the PANI chains and surface of CdS. The CdS-PANI NRs sensitized with 0.1 M $CdCl_2$, shows the lowest PL intensity and the highest peak shift, suggesting the large ϖ–ϖ* transition of the benzenoid unit and the strong chemical interaction between -NH groups of the PANI chains and surface of CdS.

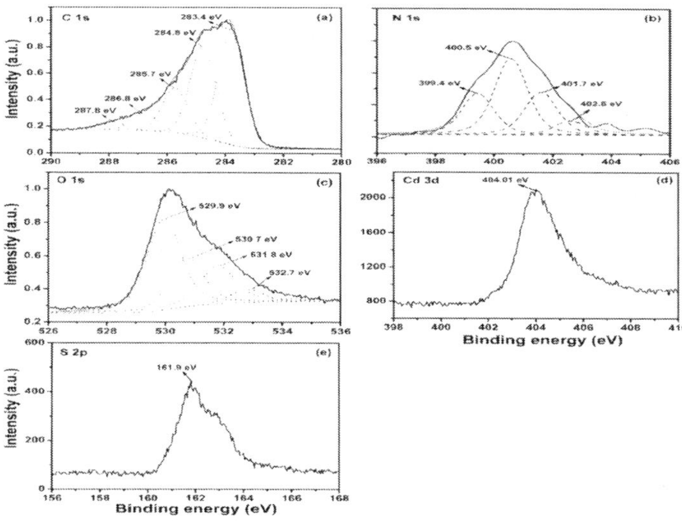

Figure 4: (a) C 1s, (b) N 1s, (c) O 1s, (d) Cd 3d and (e) S 2p XPS spectra of CdS-PANI NRs. Reprinted with permission from [Ameen S., 2012], Chem. Eng. J.181 (2012) 806© 2012, Elsevier Ltd.

The X-rays Photoelectron Spectroscopy (XPS) has studied to examine the interaction between CdS nanoparticles and the PANI NRs, as shown in Fig. 4. The C 1s XPS spectrum (Fig. 4 (a)) of CdS-PANI NRs shows the center peak at ~284.0 eV with five resolved peaks at the binding energies spanning the range from ~288 to ~283 eV. The strong peak at ~ 283.4 eV represents the carbon (C) of benzonoid ring in which a combination of protonation of imine and amine sites are formed via shake-up processes [40]. The next three resolved peaks at ~284.8 eV, ~285.7 eV and ~286.8 eV confirm the origin of the neutral C-C/C-H bond-PANI backbone, C-N^+/C=N^+ bond and C=O/C-O bond (might occur due to the absorption of moisture on the CdS-PANI), respectively [41]. The resolved peaks at ~287.8 eV assigns to the ϖ-ϖ* bonding in a long-range order with a polymer chain shake-up satellite structure and coincides with the doped states. On comparison with typical PANI peak [42], C 1s peak has shifted backwardly, suggesting that C toms of PANI is interacted with other materials (CdS, TiO_2 etc.) or impurities [43]. The O 1s XPS spectrum (Fig. 4(c)) exhibits the center peak at ~530.1 eV with three resolved peaks at ~530.7, ~531.8 and ~532.7 eV, suggesting the absorption of moisture and some oxygen impurities on the surface of CdS-PANI NRs during the synthesis. In N 1s XPS spectra (Fig. 4(b)), the main center peak with binding energy at ~400.5 eV and a resolved peak with lower binding energy at ~399.4 eV attribute to nitrogen atom originated from benzenoid diamine and quinoid di-imine nitrogen of PANI respectively. The other binding energies of ~401.7 and ~402.6 eV are ascribed to the positively charged nitrogens i.e. oxidized amine (N^+) and protonated imine (N^+) respectively [44]. The positive shifting is seen in the binding energies at ~401.7 eV and ~402.6 eV as compared with N 1s spectrum of pristine PANI indicating the involvement of positively charged nitrogen or protonated nitrogen for the partial bonding between PANI and CdS. Furthermore, the singlet peak at ~404.01 eV observes in Cd 3d XPS (Fig. 4(d)) spectrum, corresponding to Cd $3d_{5/2}$ and the typical peak of Cd^{+2} atoms in CdS [45]. Fig. 4(e)presents the S 2p XPS spectrum of CdS-PANI NRs and observed one distinct peak of $S2p_{3/2}$ at ~161.9 eV, corresponds to S^{-2} of CdS nanoparticles. This suggests the interaction and bonding between CdS nanomaterials and PANI molecules. Thus, it is concluded that the PANI and CdS nanomaterials are partially interacted and bonded by two charged nitrogen species (N^+ and N^+) of PANI with CdS nanomaterials.

Fig. 5 shows the Nyquist plot of EIS measurement for PANI NRs and CdS-PANI NRs electrodes in the electrolyte (LiI, I_2 and $LiClO_4$ in ethanol) at a frequency range from 100 kHz to 1 Hz. The almost same R_S (electrolyte resistance) with a depressed semi circle in the high frequency region is observed for the all the samples. The presence of depressed semi circle plot is ascribed to the parallel combination of the charge transfer resistance (R_{CT}) of the electrochemical reaction and the double layer capacitance (C_{dl}) of the PANI film/electrolyte interface [46] has been removed. As shown in Fig. 5, the PANI NRs electrode displays large R_{CT} value of ~17 kΩ. The R_{CT} values drastically decreases by the incremental addition of $CdCl_2$ in PANI NRs, and the order of R_{CT} values are measured as 8.4 kΩ ($CdCl_2$ (0.01 M)-PANI NRs) < 5.7 kΩ ($CdCl_2$ (0.05 M)-PANI NRs) < 4.08 kΩ ($CdCl_2$ (0.10 M)-PANI NRs). Usually, the electrode with large R_{CT} leads the slow charge transfer rate of the electrochemical system [46]. It could be seen that the highest R_{CT} value of PANI NRs electrode might result the low charge transfer rate to the electrochemical system. Moreover, the CdS decorated PANI NRs electrode with 0.01 M $CdCl_2$ presents lowest R_{CT} value which delivers the higher charge transfer rate. The variation in charge transfer resistance (R_{CT}) after the CdS sensitization by different concentrations of $CdCl_2$ might attribute to the changes in the NRs structure. Considerably, the direct band gap of CdS nanoparticles might also affect which improve the electronic state like polarons and bipolarons of PANI for the high charge carriers and enhance the charge transfer. Therefore, the CdS-PANI NRs is potential and cost effective electrode materials for the fabrication of efficient electrochemical (sensor, field emission transistor), photoelectrochemical and photovoltaic devices.

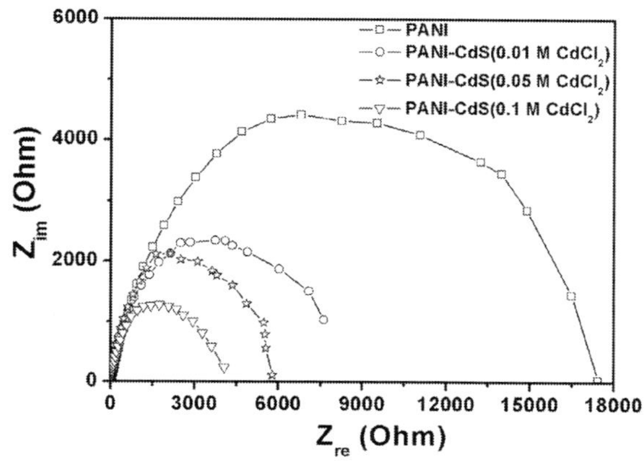

Figure 5: Nyquist plots of PANI NRs and CdS–PANI NRs at a frequency range from 100 kHz to 1 Hz. Reprinted with permission from [Ameen S., 2012], Chem. Eng. J.181 (2012) 806 ©2012, Elsevier Ltd.

BASIC STRUCTURE AND KINETICS OF DSSCS

Last few decades, considerable researches on DSSCs have extensively explored in terms of both fundamental and applied viewpoints. The basic components of DSSCs involves the conducting fluorine doped tin oxide (FTO) glass, sensitized dye, titania nanoparticles, electrolyte and Pt deposited FTO glass. The working principle of DSSC, involves the adsorption of photons by dye molecules upon light illumination and the injection of electrons from their excited states into the conduction band of the TiO_2 nanoparticles. During the entire process, the oxidized dye molecules are recharged by a redox electrolyte, which transports the positive charges by diffusion to a Pt counter electrode. The low absorption coefficient of a dye monolayer is compensated by the mesoporous structure of the TiO_2film, which leads to a strong increase in the number of TiO_2/dye/electrolyte interfaces through which photons pass, thus increases the absorption probability. The following steps are in photoelectric chemical mechanism process of DSSC:

i. $TiO_2|D + h \rightarrow TiO_2|D^*$ Excitation
ii. $TiO_2|D^* \rightarrow TiO_2|D^+ +$ (cb) Injection
iii. $TiO_2|2D^+ + 3I^- \rightarrow TiO_2|2S + I_3^-$ Regeneration
iv. $I_3^- + 2$ (Pt) $\rightarrow 3I^-$ Reduction
v. $I_3^- + 2$ (cb) $\rightarrow 3I^-$ Recaption (dark reaction)
vi. $TiO_2|S^+ +$ (cb) $\rightarrow TiO_2|S$ Recombination (dark reaction)

For the efficient working of DSSCs, the rate of re-reduction of the oxidized dye must be higher than the rate of back reaction of the injected electrons with the dye as well as the rate of reaction of injected electrons with the electron acceptor in the electrolyte. The kinetics of the reaction at the counter electrode and mesoscopic semiconductor materials with an enormous internal surface area to absorb more incident light via dye as sensitizers determines the fast regeneration of charge mediator performance [47]. Apart from this, the other significant parameters which influence the performance of DSSCs are the mesoporous morphology with high surface area of semiconducting electrode to allow absorption of a larger amount of dye and the efficient charge carrier's transport at the interface of photoanode and counter electrode by the semiconducting electrode. Moreover, the oxidized dye should be reduced to its ground state rapidly, after the injection of photoexcited electron from dye into the conduction band of semiconducting electrode. Furthermore, the semiconducting electrode must be able to permit the fast diffusion of charge carriers (higher conductivity) and produces good interfacial contact with the porous nanocrystalline layer and the counter electrode, the long-term stability, including chemical, thermal, optical, electrochemical, and interfacial stability of the electrolyte, which does not cause the desorption and degradation of the dye from the oxide surface and lastly, the optimized concentration of I^-/I_3^- which could reduce the visible light absorption by the dye, and the efficient reaction of I_3^- ions with the injected electrons to increase the dark current.

DSSCS BASED ON CONDUCTING POLYMERS

Pani as Hole Transport Materials for Dsscs

The DSSCs and polymer solar cell have been exploring the new approaches in the design of both active materials and device architectures [48]. J. Wagner et al reported the new carbazole-based polymers for DSSCs with hole-conducting polymer [49]. N. Kudo et al fabricated the organic-inorganic hybrid solar cells based on conducting polymer and SnO_2 nanoparticles which were chemically modified with a fullerene derivative [50]. S. Woo et al reported the hybrid solar cells with conducting polymers and vertically aligned silicon nanowire arrays and studied the effects of silicon conductivity [51]. F. Tan et al synthesized PANI films by electrodepostied methods and applied as anode buffer layers in solar cells [52]. M. Y. Chang et al fabricated the polymer solar cells by incorporating one-dimensional polyaniline nanotubes [53]. T. H. Lim et al utilized PANI for a flexible organic solar cells anode [54]. H. Bejbouji et al reported PANI as a hole injection layer on organic photovoltaic cells [55]. S. Zhu et al synthesized the hybrid structure of PANI/ZnO nanograss for the application in dye-sensitized solar cell1 [56]. PANI is also known as large band gap hole transporting material (HMTs) which could easily deposit as thin film on several substrates. In this regards, Ameen et al has reported the application of PANI as photoelectrode using N710 and Z907 as sensitizers for the performance of DSSCs [57].

The morphologies of ZnO nanoparticles, PANI/N719/ZnO and PANI/Z907/ZnO thin films exhibit well crystalline ZnO nanoparticles of size ~30-40 nm. The size of ZnO nanoparticles increases by ~10-15 nm from their original particle sizes after the plasma enhanced chemical vapor deposition (PECVD) polymerization of PANI molecule on dye (N719 and Z907) sensitized ZnO nanoparticulate thin film. PANI/Z907/ZnO thin film displays uniform covering or coating of PANI, indicating the well penetration of PANI on the surface of Z907/ZnO nanoparticulate thin film. The arrangement of ZnO/PANI nanoparticles are more pronounced for PANI/Z907/ZnO thin films might due to the substantive interaction and the incorporation of Z907 into ZnO nanoparticulate thin film which might allow the uniform deposition and well penetration of PANI through PECVD process. Likewise, the TEM images of ZnO nanoparticles and PANI/Z907/ZnO thin film electrode, again confirm the enhancement in the size of ZnO nanoparticles after PECVD polymerization of PANI. Significantly, after PECVD deposition of PANI, the aggregation of nanoparticles enhances the size of ZnO nanoparticles [57].

UV-Vis spectroscopy is investigated to describe the optical properties of PANI and PANI/ZnO thin film. The PECVD polymerized PANI exhibits the characteristic absorption bands at ~273 nm and ~345 nm which are ascribed to ϖ–ϖ* transitions. However, the broad band at ~611 nm is referred to *n*-ϖ* transitions which provides the information of the polarons formation into the conducting PANI. The UV-visible spectrum of PANI/ZnO thin film shows the clear red shifts with the absorption bands at ~299 nm and ~628 nm from ~273 and ~611 nm, indicating the interference in the absorption bands of PANI by ZnO nanoparticles. These shifts in the peaks are usually associated to the interactions between ZnO and PANI in PANI-ZnO thin film which might due to the existence of partial hydrogen bonding between NH (PANI).O–Zn (metal oxide) [58].

The room temperature PL spectra of PECVD polymerized PANI shows two absorption bands in blue-green region at ~438 nm and ~642 nm. The significant higher absorption wavelength shift at ~452 nm with slightly decreased peak intensity is observed in the PANI/ZnO thin film. The considerable changes in the PL emission peak might arise due to the effective interaction between imine (-NH) of PANI and hydroxyl (-OH) group of ZnO nanoparicles in PECVD polymerized PANI-ZnO thin film [57].

The current density-voltage (J-V) characteristics at one sun light illumination (100 mW/cm^2, 1.5AM) have been carried out to evaluate the performances of solar cell fabricated with PANI/N719/ZnO and PANI/Z907/ZnO thin film electrode. The measurements of V_{OC}, J_{SC}, FF and overall solar-to-electrical energy conversion efficiency are obtained from the J-V characteristics of both the DSSCs. The PANI/N719/ZnO electrode based DSSC exhibits low solar-to-electricity conversion efficiency of ~0.6%, with J_{SC} of ~2.80 mA/cm^2, V_{OC} ~0.432 V and FF ~0.51, whereas, PANI/Z907/ZnO electrode based DSSC executes the greater overall solar-to-electricity conversion efficiency of ~1.31% with $V_{OC}$of J_{SC} ~4.56 mA/cm^2, ~0.521 V, and FF ~0.55. On comparison with PANI/N719/ZnO electrode, DSSC with PANI/Z907/ZnO electrode attains considerably improved the solar-to-electricity conversion efficiency by ~53% along with other parameters of J-V characteristics. The enhanced performances and J_{SC} might attribute to the fast movements of photon generated electrons at the interface of the PANI/ZnO and the nature of Ru dye (Z907) with long chain alkyl group at pyridine rings. Moreover, as seen in FESEM results, the high penetration of the hole conductor (PANI) into the pores of Z907 sensitized ZnO thin film might execute reasonably fast charge injection and electron transfer at the interface of PANI and ZnO layer to Pt layer of electrode. Thus, the choice of dye is crucial to obtain the high performance DSSC with PECVD polymerized PANI/ZnO electrodes. In other previous report, Ameen et al showed the effects of PANI on TiO_2 as an effective photoelectrode for the performance of DSSCs [59].

A schematic energy level diagram for the device FTO/TiO_2/Dye/PANI/Pt is shown in Fig. 6 (a). The diagram, in accordance with step (1) indicates that the electrons from the dye, upon illumination, jumps from the HOMO level to the LUMO level and thus, as per step (2), these electrons are transferred from the conduction band (C.B) to the valence band (V.B) of TiO_2. The step (3) shows the transfer of electrons from V.B to the HOMO of PANI. As indicated by step (4), further transfer of electrons could be preceded through two different possible ways. Firstly, the electron could move either through the LUMO of PANI, followed by step (5) or then may jump to the LUMO of the dye and finally move onwards by repeating the step (2). Secondly, electrons, from step (3), could also jump to the HOMO of dye and would move ahead by following the same step (1), leading to the transfer of electrons to the entire cell. During the entire cycle, the recovery of the holes is accomplished at the counter electrode. Additionally, Fig. 6(b) depicts that PANI participates in the light absorption through the effective injection of electrons from its LUMO to the C.B of TiO_2. Therefore, the proposed mechanism presents that FTO/TiO_2/Dye/PANI/Pt system might deliver the high transportation of charge carriers during the operation of device under the illumination.

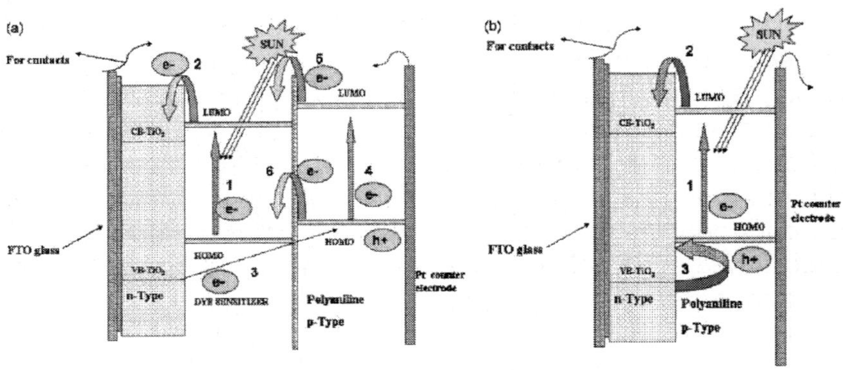

Figure 6: An overview of the energy level diagram of the fabricated devices (a) FTO/TiO_2/Dye/PANI/Pt (b) FTO/TiO_2//PANI/Pt. Reprinted with permission from [Ameen S., 2009], J. Alloys comp. 487 (2009) 382.© 2011, Elsevier Ltd.

To elucidate the charge transfer properties of TiO_2/PANI electrodes, an electrical impedance spectroscopy (EIS) measurement is used. According to the diffusion recombination model proposed by Bisquert [60], an equivalent circuit representing device is illustrated [(Inset of Fig. 7 (a)]. Equivalent circuit is composed of the series resistance (R_S), the charge transfer resistance at the junction of $TiO_2$and PANI layer in TiO_2/PANI or TiO_2/Dye/PANI electrodes (R_{CT}), the charge transfer resistance at the interface of TiO_2/PANI or TiO_2/Dye/PANI and TCO ($R_{P/TCO}$), the capacitance of accumulation (of e^-) layer of the TiO_2 (C_{acc}), and space charge capacitance ($C_{SC)}$ [61]. The values of real impedance (Z_{re}) are used to estimate the values of $R_{P/TCO}$ and R_{CT} at different frequencies. Fig. 7 (a, b) exhibits the Nyquist curve of cell fabricated with TiO_2/PANI and TiO_2/Dye/PANI electrodes. A very high $R_{P/TCO}$ of

52.4Ω and R_{CT} of 3700 Ω observed for TiO_2/PANI electrodes based cells, which are estimated from Fig. 7(a). Comparatively, TiO_2/Dye/PANI based device (Fig.7 (b)) shows the low $R_{P/TCO}$ (35.8 Ω) and R_{CT} (81.9 Ω) due to the influence of dye layer which is placed between the TiO_2 and PANI layer of the electrode. It is reported that a small R_{CT} of the device suggests the fast electron transfer, while a large R_{CT} indicates the slow charge transfer at the junction of inorganic and organic layer [62]. In our case, it is found that the value of R_{CT} in TiO_2/Dye/PANI based device is very low as compared to the R_{CT} of TiO_2/PANI based device. Therefore, it explains the high electron transfer at the junction of TiO_2 and PANI in TiO_2/Dye/PANI based device, resulting in the high photocurrent density and overall conversion efficiency, which are in the excellent agreement with the J-V curve results of the devices. The impedance results clearly indicate that the high photocurrent density, high overall conversion efficiency and low R_{CT} are resulted from the uniform distribution of PANI molecules on the mesoporous surface of TiO_2 electrode. Therefore, the lower R_{CT} and $R_{P/TCO}$ in TiO_2/Dye/PANI based device reveals that the dye and PANI layers on the surface of TiO_2 electrode play an important role in the charge transfer at hole conductor (PANI)-dye absorbed TiO_2 region, which results the high J_{SC}, FF, and conversion efficiency than that of TiO_2/PANI electrode based cells.

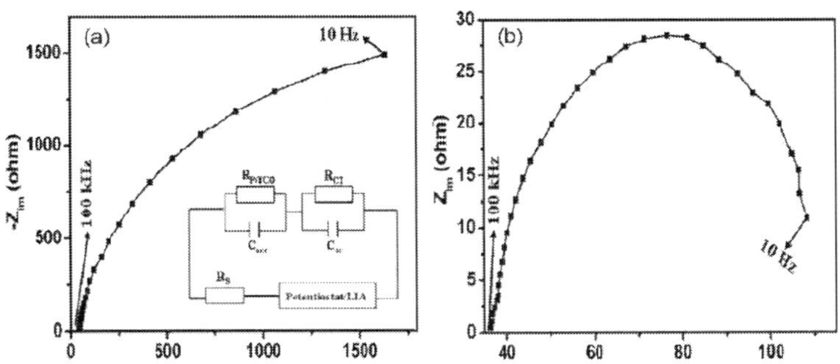

Figure 7: AC impedance of (a) FTO/TiO_2/PANI/Pt and (b) TiO_2/Dye/PANI thin film electrode based DSSCs at the frequency range from 10 Hz to 100 kHz. Inset shows the equivalent circuit model of the device. Reprinted with permission from [Ameen S., 2009], J. Alloys comp. 487 (2009) 382. ©2011, Elsevier Ltd.

The J-V performance of solar cell FTO/TiO_2/Dye/PANI/Pt and FTO/TiO_2/PANI/Pt are shown in Fig. 8 (a, b) under 100 mW/cm² light intensity. On comparison with TiO_2/PANI, the solar cell based on TiO_2/Dye/PANI electrode executes great improvement in the overall conversion efficiency with the incorporation of dye layer on TiO_2/PANI electrode. The conversion efficiency of the solar cell drastically increases from ~0.005% to ~0.68%. It is noteworthy that the photovoltaic properties such as V_{OC}, J_{SC} and FF of the DSSCs enhance dramatically as compared to TiO_2/PANI electrode based DSSC. The high J_{SC} is imputed to the high electrical conductivity of PANI/TiO_2 thin film. The enhancement in JV parameters are resulted from the formation of TiO_2/PANI thin film, where the photon generated electrons could freely travel at the interface of PANI and TiO_2 without decay, and dissociate into free charge carriers effectively. Moreover, the pore filling extent of the hole conductor into the dye-sensitized TiO_2 film, and the electric contact of the hole conductor are the two important factors to determine the photovoltaic behaviors of device. The advanced TiO_2/Dye/PANI electrode executes reasonably fast charge injection and transfer of electron at the interface of hole conductor (PANI) and Pt layer of electrode.

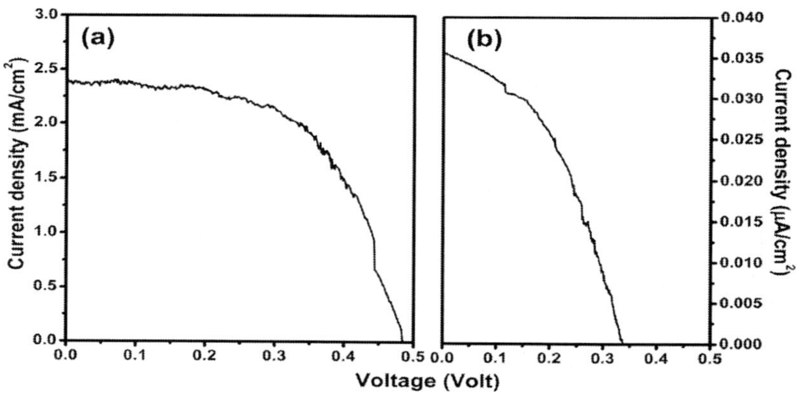

Figure 8: J–V curve of fabricated solar cell (a) FTO/TiO_2/Dye/PANI/Pt (b) FTO/TiO_2//PANI/Pt. Reprinted with permission from [Ameen S., 2009], J. Alloys comp. 487 (2009) 382 ©2011, Elsevier Ltd.

Pani as Counter Electrodes for Dsscs

The counter electrode in DSSCs is responsible for the electrocatalytic reduction of I_3^- ions. Until now, Pt counter electrode shows the high electrocatalytic activity for I_3^- ions reduction, high conductivity, and stability. Pt is one of the most expensive components in DSSCs. Therefore, the development of counter electrodes with alternative materials is expected to reduce production costs of DSSCs. Several varieties of materials such as carbon nanotubes, activated carbon, graphite, and conducting polymers are employed as active catalysts for counter electrodes. In this regards, M. H. Yeh et al reported the conducting polymer-based counter electrode for a quantum-dot-sensitized solar cell (QDSSC) with a polysulfide electrolyte [63]. K. M. Lee et al fabricated the DSSC based on poly (3, 4-alkylenedioxythiophene) as counter electrode [64]. In another report, K. M. Lee et al exhibited the effects of mesoscopic poly (3, 4-ethylenedioxythiophene) films as counter electrodes for DSSCs [65]. W. Maiaugree et al optimized the TiO_2 nanoparticle mixed PEDOT-PSS counter electrodes for high efficiency of DSSCs [66]. J. Chen et al reported polyaniline nanofiber-carbon film as flexible counter electrodes in platinum-free dye-sensitized solar cells [67]. Q. Li et al fabricated the microporous polyaniline thin film as counter electrode for DSSCs [68]. J. Zhang et al applied the nanostructured PANI thin film as counter electrode for DSSCs and investigated the electrochemical formation mechanism [69]. Furthermore, G. Wang et al synthesized PANI-graphene hybrids thin film and utilized as a counter electrode in DSSCs [70]. Tai et al prepared the highly uniform and transparent PANI counter electrodes by a facile in situ polymerization method for the DSSCs [71]. In this regards, Ameen et al performed the doping of PANI with sulfamic acid (SFA) and applied as counter electrode for the efficient performance of DSSCs [72].

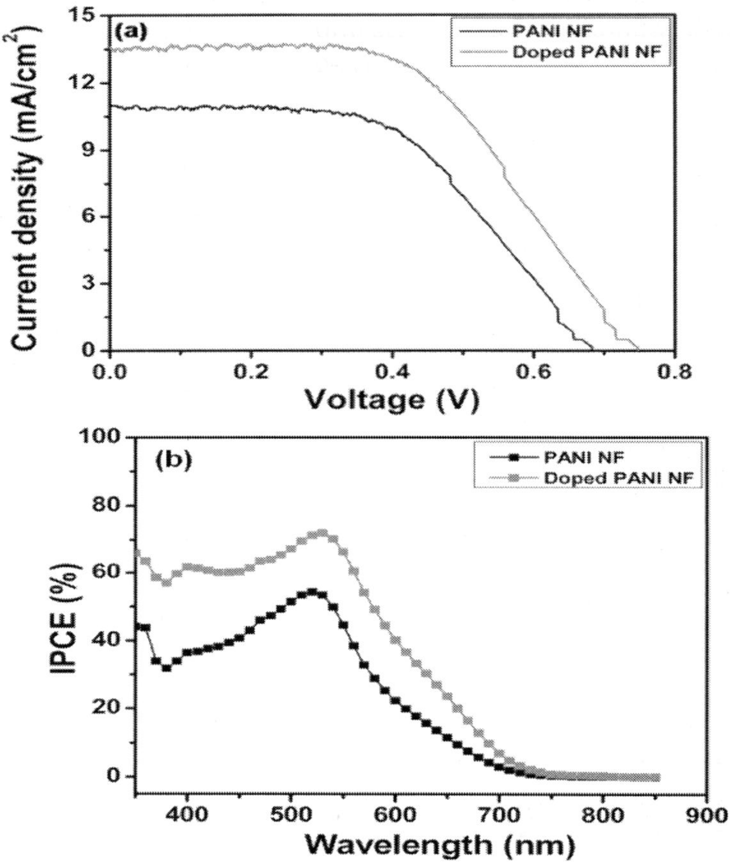

Figure 9: J-V curve (a) and IPCE curves (b) of a fabricated solar cell of PANI NFs and SFA doped PANI NFs as counter electrodes. Reprinted with permission from [Ameen S., 2010], J. Phys. Chem. C 114 (2010) 4760. ©2010, American Chemical Society.

Fig. 9(a) shows the J-V curve of the DSSCs fabricated with the counter electrodes made of PANI NFs and SFA-doped PANI NFs under dark and light intensity of 100 mW/cm^2 (1.5AM). DSSCs fabricated with SFA-doped PANI NFs counter electrode achieve high conversion efficiency ( ) of ~5.5% with J_{SC} of ~13.6 mA/cm^2, V_{OC} of ~0.74 V, and FF of ~0.53. Significantly, the conversion efficiency increases from ~4.0% to ~5.5% after SFA doping into the PANI NFs. DSSC fabricated with SFA-doped PANI NFs counter electrode has appreciably improved the conversion efficiency and J_{SC} by ~27% and ~20% than that of DSSC fabricated with PANI NFs counter electrode. These improvements are resulted from the higher electrocatalytic activity of SFA-doped PANI NFs, which serves a good path for the charge transport of I^-/I_3^- redox. Therefore, the superior photovoltaic properties such as , J_{SC}, and V_{OC} of the cell are attributed to the sufficiently high conductivity and electrocatalytic activity of doped PANI NFs, which alleviates the reduction of I_3^- at the thin SFA-doped PANI NFs layers. Fig. 9(b) presents the IPCE curves of DSSCs fabricated with PANI NFs and SFA-doped PANI NFs counter electrodes. DSSCs fabricated with PANI NFs counter electrode exhibits the low IPCE of ~54% in the absorption range of 400-650 nm. The IPCE value is prominently increased by ~70% with the SFA doped PANI NFs counter electrode-based DSSCs. It is noteworthy that the IPCE of the device is considerably enhanced by ~24% upon SFA doping on PANI NFs-based counter electrodes. The enhanced IPCE results are consistent with high electrical conductivity [73] and the electrocatalytic activity of the SFA-doped PANI NFs electrode. Thus, SFA doping significantly enhances the electrical conductivity and increases the higher reduction of I_3^- to I^- in the electrolyte at the interface of PANI NFs layer and electrolyte.

Other Ions Doped Pani Counter Electrode Based Dsscs

Z. Li et al recently studied on the in situ electropolymerized-PANI thin film of thickness ~5–20 μm, deposited on FTO glass. The PANI thin films were doped by various counter ions like SO_2^{-4}, ClO^{-4}, BF^{-4}, Cl^-, p-toluenesulfonate (TsO^-), etc. Different doping counter ions showed different impact on the morphology, electrochemical activity of the electropolymerized-PANI thin film. The electropolymerized-PANI doped by SO_2^{-4} anion (PANI-SO_4) film was much porous morphology with pore size diameter of several micrometers and possessed the higher reduction current for the reduction of I_3^- and a low charge transfer resistance of 1.3 Ωcm^2 as compared with Pt as counter electrode (CE). Dye-sensitized solar cell (DSSC) with PANI-SO_4 as CE was assembled, and the device under full sunlight illumination (100mWcm^{-2}, AM 1.5 G) showed ~5.6% photovoltaic conversion efficiency, which was comparable to ~6.0% of that with Pt CE under the same experimental condition. The electropolymerized-PANI doped with SO_2^{-4} ion with a porous and homogeneously structure was a promising candidate which showed the high performance of DSSC. On the other hand, PANI-BF_4 and PANI-Cl was porous and fibrillar thin film which exhibited the modest efficiency of ~3.9% and ~2.6%. On the other hand, PANI-ClO_4 and PANI-TsO showed the very low performance of DSSC ca. <1%, and the R_{CT} was greatly increased accordingly to over 100 Ωcm^2 [74].

DSSCS BASED ON METAL OXIDE SEMICONDUCTORS

In DSSCs, the choice of semiconductor is governed by the conduction band energy and density of states, which facilitate the charge separation and minimize the recombination. Secondly, the high surface area and morphology of semiconductors is important to maximize the light absorption by the dye molecules while maintaining the good electrical connectivity with the substrate [75]. The semiconducting metal oxides such as titania (TiO_2), zinc oxide (ZnO), and tin oxide (SnO_2) have shown good optical and electronic properties and are accepted as the effective photoelectrode materials for DSSCs. To improve the light harvesting efficiency, the metal oxide nanostructures must possess a high surface-to-volume ratio for high absorption of dye molecules. Particularly in DSSCs, the porous nature of nanocrystalline TiO_2 films provides the large surface for dye-molecule adsorption and therefore, the suitable energy levels at the semiconductor-dye interface (the position of the conduction band of TiO_2 being lower than the excited-state energy level of the dye) allow for the effective injection of electrons from the dye molecules to the semiconductor. Compared with other photovoltaic materials, anatase phase TiO_2 is outstanding for its stability and wide band gap and, thus, widely used in the devices [76]. On the other hand, ZnO nanomaterials are chosen as an alternative material to TiO_2 photoanodes due to their wide band gap with higher electronic mobility, which would be favorable for the efficient electron transport, with reduced recombination loss in DSSCs. Studies have already been reported on the use of ZnO material photoanodes for the application in DSSCs. Although the conversion efficiencies of ZnO (0.4-5.8%) is comparably lower than TiO_2 (~11%), ZnO is still a distinguished alternative to TiO_2 due to its ease of crystallization and anisotropic growth. In this part of the chapter, the TiO_2 and ZnO have been briefly summarized for the application for DSSCs.

Dsscs Based On Tio_2 Photoanode

Due to versatile and the exotic properties, TiO_2 nanomaterials are so far used in many technological applications as a photocatalyst, photovoltaic material, gas sensor, optical coating, structural ceramic, electrical circuit varistor, biocompatible material for bone implants, and spacer material for magnetic spin valve systems etc [77]. The dimensionality of TiO_2 at the nanoscale level is the crucial characteristic for determining the physiological and electrical properties. In recent years, one dimensional (1D) TiO_2 nanomaterials like NRs, NWs and NTs display significantly larger surface areas as compared to bulk materials, which deliver unique chemical and the physical properties [78] and contribute towards the electrical and photoelectrochemical applications [79]. The 1D TiO_2 such as NRs and NTs have shown the reduced recombination rate for the excited electron-hole pair and display unique optical and the electric properties [80]. Particularly, the vertically grown TiO_2 NRs allow shorter and the uninterrupted electrical pathways for the photogenerated carriers and improves the charge separation and charge transport properties in many photoelectrochemical devices like dye sensitized solar cells (DSSCs) [81]. As compared to TiO_2 nanoparticles, it is expected that the highly oriented TiO_2 NRs could be the potential electrode and photocatalyst material for several photoelectrochemical applications. S. Ameen et al reported the TiO_2 nanorods (NRs) based photoanode for the fabrication of DSSCs [82].

The morphology of the TiO_2 NR thin films deposited on FTO substrates by the hydrothermal process with variations of the ethanol/DI water precursor solution is shown in FESEM images (Fig. 10). With an ethanol/DI water ratio of 0:100 v/v as the precursor solution, the distorted hexagonal TiO_2 NRs (Fig. 10 (a, b)) of average diameter ~100-200 nm and length of 3.0 mm are obtained. However, the round headed TiO_2 NRs with ethanol/DI water (50: 50 v/v) as the precursor solution (Fig. 10 (c, d)) is formed. The highly ordered tetragonal TiO_2 NRs have grown on the FTO substrate with the precursor solution of ethanol/DI water (80: 20 v/v) as seen in Fig. 10 (e, f). The grown TiO_2 NRs possess the average lengths of 2-4 mm and diameters of ~50–70 nm respectively. The high amount of ethanol in the precursor solution is crucial to achieve the highly ordered nanorods.

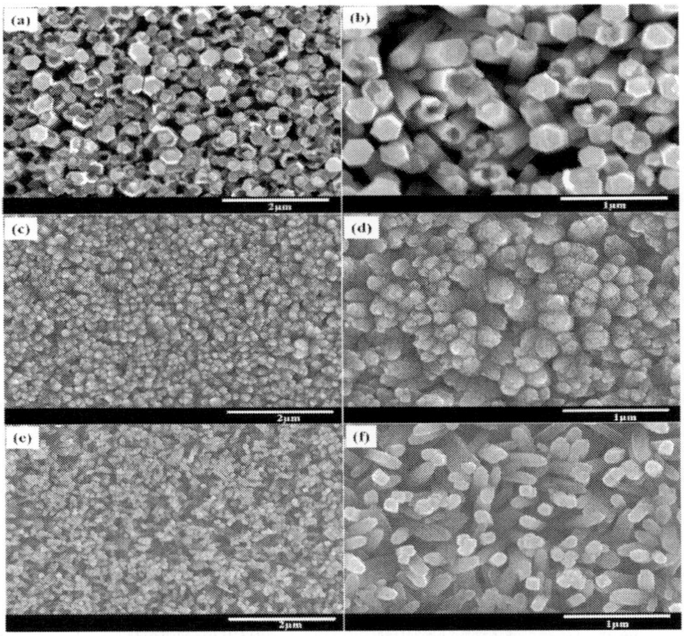

Figure 10: Low and high resolution FESEM images of the TiO_2 NR thin films obtained with the precursor solutions of ethanol/DI water with ratios of (a, b) 0 : 100 v/v, (c, d) 50 : 50 v/v and (e, f) 80 : 20 v/v. Reprinted with permission from [Ameen S., 2012], RSC Adv. 2 (2012) 4807 ©2012, RSC Publications Ltd.

Fig. 11 shows the transmission electron microscopy (TEM), high resolution (HR) TEM and the selected area electron patterns (SAED) of the grown TiO_2 NR coated FTO substrate. Similar to the FESEM results, the highly ordered tetrgonal TiO_2 NRs from the precursor solution of ethanol/DI water (80:20v/v) solvent comprises the average length of 2-4 mm and the diameter of 50-70 nm, as shown in Fig. 11 (a). Each NR is made of a bundle of the densely packed nanofibers (NFs) with an average fibril's diameter of ~5 nm. The corresponding SAED pattern (Fig. 11 (b)) displays the clear phases, suggesting the high crystal quality with the single crystalline fibrils derived from TiO_2 NRs along the [001] direction. However, the HRTEM image (Fig. 11 (c)) shows the well-resolved lattice fringes of the grown TiO_2 NRs and estimates an average interplanar distance of 0.35 nm between the two fringes, which reveals the typical interplanar distance of anatase TiO_2 [83]. On the other side, the width and length of distorted hexagonal TiO_2 NRs are respectively observed as ~200 nm and ~3.2 mm, as seen in Fig. 11(d).

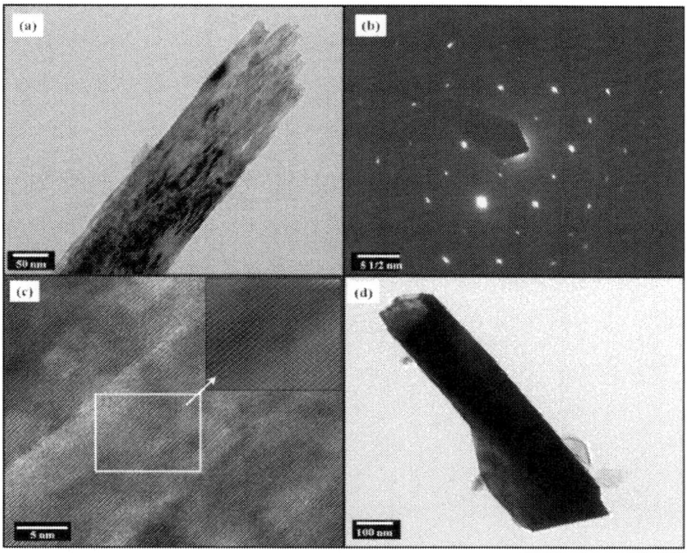

Figure 11: TEM image of (a) highly ordered tetragonal TiO_2 NRs, (b) SAED patterns, (c) HRTEM image and (d) TEM image of grown hexagonal distorted TiO_2 NRs. Reprinted with permission from [Ameen S., 2012],], RSC Adv. 2 (2012) 4807 ©2012, RSC Publications Ltd.

The XRD patterns (Fig. 12) of grown TiO_2 NRs from both precursor solutions exhibit the anatase phase with the peaks at 25.1°, 37.9°, 48.1°, 53.8° and 55.1°, which correspond to typical anatase TiO_2materials and indexes at JCPDS no. 89-4203. However, the diffraction peaks of the FTO substrate are also observed at 33.8°, 35.7° and 52.8 ° (JCPDS no. 88-0287). On comparison with the distorted hexagonal TiO_2 NRs, the intensities of XRD diffraction peaks have slightly changed, which might indicate the high crystalline nature of the highly ordered tetragonal TiO_2 NRs.

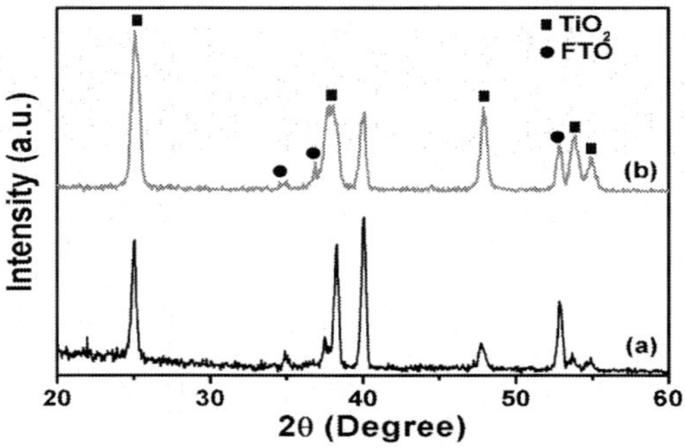

Figure 12: XRD patterns of (a) the distorted hexagonal TiO_2 NRs and (b) the highly ordered tetragonal TiO_2NR thin film. Reprinted with permission from [Ameen S., 2012], RSC Adv. 2 (2012) 4807 ©2012, RSC Publications Ltd.

The J-V characteristics (Fig. 13 (A)) have been performed to elucidate the performance of the DSSCs fabricated with the photoanodes of grown TiO_2 NRs and are measured under a light intensity of 100 mW cm^{-2} (1.5 AM). DSSC fabricated with the distorted hexagonal TiO_2 NRs photoanode shows a relatively low solar efficiency of ~1.08%, with a low J_{SC} of ~4.48 mA cm^{-2}, V_{OC} ~0.571 V and FF of ~0.42. However, DSSCs fabricated with the highly ordered tetragonal TiO_2 NRs photoanode achieves an appreciably improved overall conversion efficiency of ~3.2% with a high J_{SC} of ~8.7 mA cm^{-2}, V_{OC} of ~0.67 V, and FF of ~0.54. As compared to the distorted hexagonal TiO_2 NRs photoanode based DSSC, the photovoltaic performance, J_{SC}, V_{OC} and FF are significantly enhanced by ~ 67%, ~48%, ~15% and ~22% respectively. It is seen that the size of the NRs also plays an important role for achieving the high photocurrent density and performance of the device. It is known that the high photovoltaic performance and photocurrent density are related to high light harvesting through the highly uniform and high surface to volume ratio of the photoanode materials [84]. In general, the TiO_2 thin film electrodes with larger particles have the smaller surface area and produce moderate contact points between nanoparticles at the interface of the sintered nanoparticles and the conducting substrate, leading to the lower availability of the active surface for dye adsorption, which perhaps decreases the amount of light absorbed and generates the large number of electrons and holes. Whereas, the TiO_2 thin film with smaller particles acquires the larger surface area and higher number of contact points of the sintered colloidal particles present at the interface of the nanoparticles and the conducting substrate, which gives rise to larger dye adsorption and higher light harvesting efficiency [85]. In this case, the distorted hexagonal TiO_2 NRs consist of larger diameters and lengths as compared to the highly ordered TiO_2 NRs, as shown in the FESEM images. It is believed that the smaller diameters of the NRs might generate the high light harvesting efficiency, which might lead to the high photocurrent density and the conversion efficiency. From the UV-Vis spectra (Fig. 13 (B)) of the dye desorption from dye absorbed TiO_2 NRs photoanodes in NaOH solution, the photoanode of highly ordered tetragonal TiO_2 NRs attains the higher dye loading than the photoanode of the distorted hexagonal TiO_2 NRs. Herein, the enhanced photovoltaic performance and J_{SC} are related to the highly ordered NRs morphology, high dye loading and improved light harvesting efficiency through the high surface area of the film. Besides these, the unique ordered morphology of the NRs might retard the recombination rate and contribute to longer electron lifetimes [86], resulting in the increased V_{OC} and FF of device.

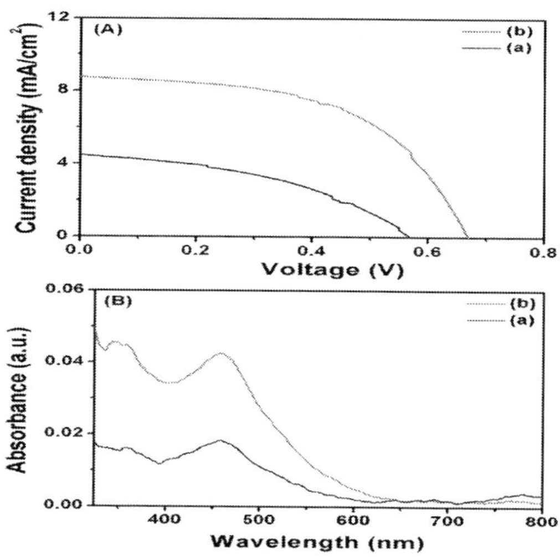

Figure 13: (A) J-V curve of the DSSC fabricated with (a) distorted hexagonal TiO_2 NRs and (b) highly ordered tetragonal TiO_2 NRs. (B) UV-Vis spectroscopy of desorbed dye from (a) distorted hexagonal TiO_2NRs and (b) highly ordered tetragonal TiO_2 NRs. Reprinted with permission from [Ameen S., 2012]], RSC Adv. 2 (2012) 4807©2012, RSC Publications Ltd.

The IPCE (Fig. 14) of DSSCs fabricated with highly ordered tetragonal TiO_2 NRs and distorted hexagonal TiO_2 NRs photoanodes have shown the broad the absorption edge of visible spectrum from 400-800 nm. The highly ordered tetragonal TiO_2 NRs photoanode based DSSC exhibits the maximum IPCE of ~31.5% at the highest absorption edge of ~528 nm, whereas ~17.9% IPCE at ~528 nm is achieved by the distorted hexagonal TiO_2 NRs photoanode based DSSC. The highly ordered tetragonal TiO_2 NRs photoanode based DSSC considerably improves IPCE by approximately two times to DSSC with distorted hexagonal TiO_2 NRs photoanode, which is attributed to the high dye loading of the photoanode, resulting in the high light harvesting efficiency and the electron injection from dye to CB of TiO_2. Thus, the highly ordered tetragonal TiO_2 NRs photoanode with enhanced dye loading, light harvesting and IPCE, have resulted to increased J_{SC}, V_{OC} and the photovoltaic performance for DSSC.

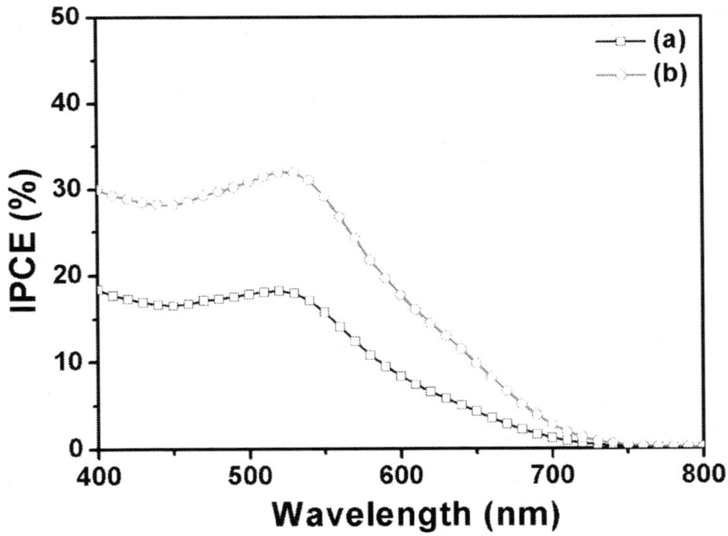

Figure 14: IPCE curve of the DSSC fabricated with (a) the distorted hexagonal TiO_2 NR photoanode and (b) the highly ordered tetragonal TiO_2 NR photoanode. Reprinted with permission from [Ameen S., 2012], RSC Adv. 2 (2012) 4807©2012, RSC Publications Ltd.

Dsscs Based Zno Photoanode

Figure 15: FE-SEM images of (a) ZnO–NH3, (b) ZnO-citric, and (c) ZnO-oxalic materials prepared by hydrothermal method. Inset shows the high magnification FE-SEM images of (a), (b), and (c). Reprinted with permission from [Akhtar., 2008], Electrochim. Acta 53(2008)7869.© 2008, Elsevier Ltd.

Another metal oxide nanomaterials such as ZnO nanomaterials, are recently dealing with the versatile applications in various fields such as field-effect transistors, lasers, photodiodes, sensors and photovoltaics owing to their unique photoelectric properties, optical transparency, electric conductivity and piezoelectricity properties [87]. Importantly, ZnO nanomaterials possess similar wide band gap (~3.37 eV) with large exciton binding energy (~60 meV) and higher electron mobility [88]. Moreover, ZnO nanomaterials with different nanostructures have presented the versatile properties like higher surface-to-volume ratio, chemical stability, high exciton binding energy, and moderate charge transport capability [89] and therefore, it becomes one of the promising alternatives of nanocrystalline TiO_2 photoanode in DSSCs and hybrid solar devices. To control the parameters like morphology, physical and the crystalline properties of ZnO nanomaterials, the performance of DSSCs could be accelerated [90]. So far, the different morphologies of ZnO nanostructures such as nanorods [91], nanotetrapods, nanosheet [92] and nanobelts based photoanodes have been studied for the fabrication of efficient DSSCs [93] and achieved encouraging results. Law et al. firstly designed ZnO NW arrays to increase the electron diffusion length, and applied as photoelectrodes for the fabrication of DSSCs [94]. The grown ZnO NW array films exhibited relatively good resistivity values that ranged from 0.3 to 2.0 Ωcm for individual NWs and a mobility of 1-5 $cm^2 V^{-1}s^{-1}$. The overall conversion efficiencies of ~1.2-1.5% were obtained by DSSCs fabricated with ZnO NW arrays with J_{SC} of ~5.3-5.85 mA/cm^2, V_{OC} of ~0.610-0.710 V, and FF of ~0.36-0.38. Another group synthesized ZnO NWs by the use of ammonium hydroxide for changing the super saturation degree of Zn precursors in solution process [95]. The length-to-diameter aspect ratio of the individual NWs was easily controlled by changing the concentration of ammonium hydroxide. The fabricated DSSCs exhibited remarkably high conversion efficiency of ~1.7%, which was much higher than DSSC with ZnO NR arrays [96]. Jiang et al. fabricated the flexible DSSCs with a highly bendable ZnO NWs film on PET/ITO substrate which was prepared by a low-temperature hydrothermal growth at 85°C [97]. The fabricated composite films obtained by immersing the ZnO NPs powder in a methanolic solution of 2% titanium isopropoxide and 0.02 M acetic acid was treated with heat, which favored a good attachment of NPs onto the NW surfaces [97]. Here, the achieved conversion efficiency of the fabricated DSSCs was less as compared to DSSCs based on NPs. Recently, Akhtar et al. demonstrated that the performance of DSSCs effectively altered by varying the morphologies of ZnO nanomaterials. They reported the morphology of ZnO nanomaterials through a hydrothermal process using Zinc acetate, NaOH and different capping agents, as shown in Fig 15. The photoanode was prepared by spreading the ZnO paste on an FTO substrate by a doctor blade technique for the fabrication of DSSCs [98], and they obtained non-uniform surface of film. Unfortunately, the DSSCs with ZnO NRs photoanode presented a very low conversion efficiency of ~0.3% with a high FF of ~0.54 (Fig 16), which might attribute to the low dye absorption on the surface of ZnO NRs due to the less uniformity of the thin film with low surface-to-volume ratio (Fig 17). Furthermore, a flower-like structure consisted of NRs/NWs could deliver a larger surface area and a direct pathway for electron transport with the channels arisen from the branched to NR/NW backbone. Recently, hydrothermally grown ZnO nanoflower films accomplished an improved overall conversion efficiency of ~1.9%, with a high J_{SC} of ~5.5 mA/cm^{2}and an FF of ~0.53 [99]. These parameters are higher than NR arrays film-based DSSCs of the conversion efficiency ~1.0%, J_{SC} ~4.5 mA/cm^2, and FF ~0.36. Recently, S. Ameen et al reported the nanospikes decorated ZnO sheets thin film as photoanode for the performance of DSSCs [100].

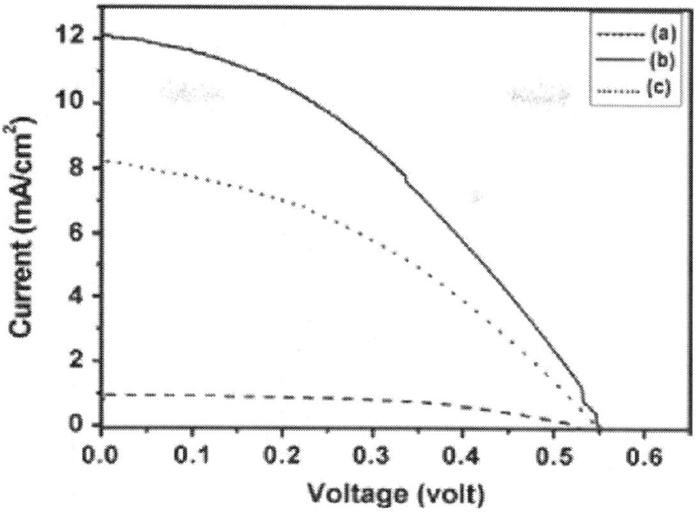

Figure 16: Current–voltage curve of DSSC fabricated with (a) ZnO–NH_3, (b) ZnO-citric, and (c) ZnO-oxalic at 1.5 AM. Reprinted with permission from [Akhtar., 2008], Electrochim. Acta 53, (2008) 7869.© 2008, Elsevier Ltd.

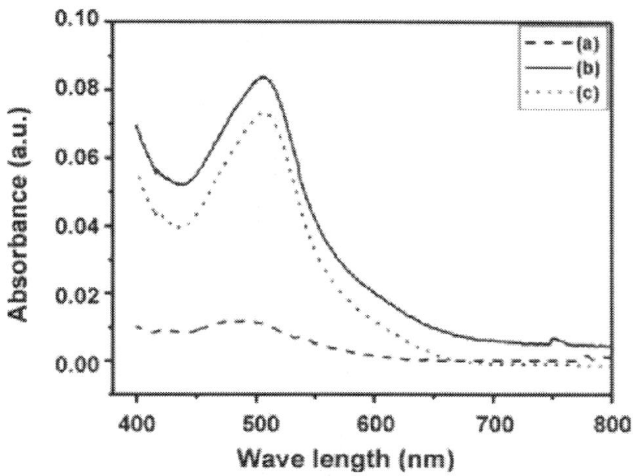

Figure 17: UV-vis absorption of dye (N-719) extracted with 2 M NaOH from the electrodes of (a) ZnO–NH_3, (b) ZnO-citric, and (c) ZnO-oxalic. Reprinted with permission from [Akhtar., 2008], Electrochim. Acta 53 (2008) 7869.© 2008, Elsevier Ltd.

The FESEM image (Fig. 18) shows dense and uniform deposition of the nanospikes decorated ZnO sheets morphology on the FTO substrate. Each nanospikes decorated ZnO sheets is comprised of a sheet with the average thickness of ~50-60 nm and the aligned nanospikes with the average diameter of ~80-100 nm and length of ~150-200 nm. Interestingly, the nanospikes are consisted of the bundles of small nanorods. The nanospikes are aligned either on one side or other side of ZnO sheet, but in some cases, these nanospikes are found on the both sides of ZnO sheets.

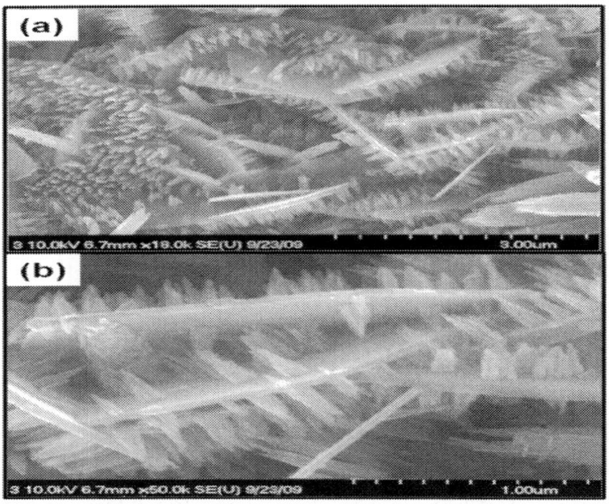

Figure 18: FESEM images of nanospikes decorated ZnO sheets (a) at low magnification and (b) at high magnification. [Ameen S., 2012], Chem. Eng. J. 195 (2012) 307©2012. Elsevier Ltd.

Similarly, TEM images (Fig. 19(a)) present the nanospikes with nanosheets in which nanospikes decorated on both the sides of ZnO sheet. The average thickness of the sheet is ~50-60 nm and the decorated nanospikes possess the average diameter of ~30 nm (single rods) and the length of ~150-200 nm. From the HRTEM image (Fig. 19 (b)), the well-resolved lattice reveals that the grown ZnO nanomaterials exhibit the good crystallinity. The inter-planar spacing of ~0.52 nm is observed which is consistent to the lattice constant in the reference (JCPDS No. 36-1451) for ZnO nanomaterials. This inter-planar spacing value of the lattice fringes correspond to the [0001] crystal plane of the wurtzite ZnO confirms that the grown ZnO nanomaterials are almost defect free [100]. Moreover, the corresponding selected area electron diffraction (SAED) also indicates the typical wurtzite single crystalline structure and the ZnO nanomaterials are grown along caxis direction [0001].

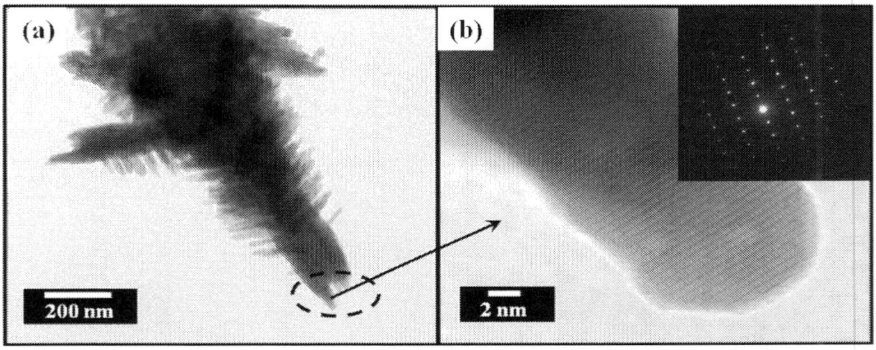

Figure 19: TEM image (a) and HR-TEM image (b) of grown nanospikes decorated ZnO sheets. Inset shows the corresponding SAED patterns of grown nanospikes decorated ZnO sheets. Reprinted with permission from [Ameen S., 2012], Chem. Eng. J. 195 (2012) 307 ©2012, Elsevier Ltd.

XRD patterns (Fig. 20 (a)) of the nanospikes decorated ZnO sheets obtain all the diffraction peaks appeared at 32.3° (100), 35.2° (002), 36.8° (101), 48.2° (102), 57.2° (110), 63.5° (103) and 66.2°(200) which are well matched with the JCPDS card No.36-1451. It confirms that the ZnO nanomaterials possess the hexagonal wurtzite phase with the lattice parameters: a-3.246 and c-5.206 Å. The intensity of (101) diffraction peak is much higher compared to other peaks, indicating the preferential growth direction due to the instability of polar (101) plane [101]. A single narrow absorption peak is observed near the UV region at ~376 nm in the UV-Vis absorbance spectrum of nanospikes decorated ZnO sheets structures (Fig. 20 (b)), corresponds to the characteristic band of the wurtzite hexagonal structure in bulk ZnO [102]. Moreover, the single peak suggests purity of the grown nanospikes decorated sheets morphology.

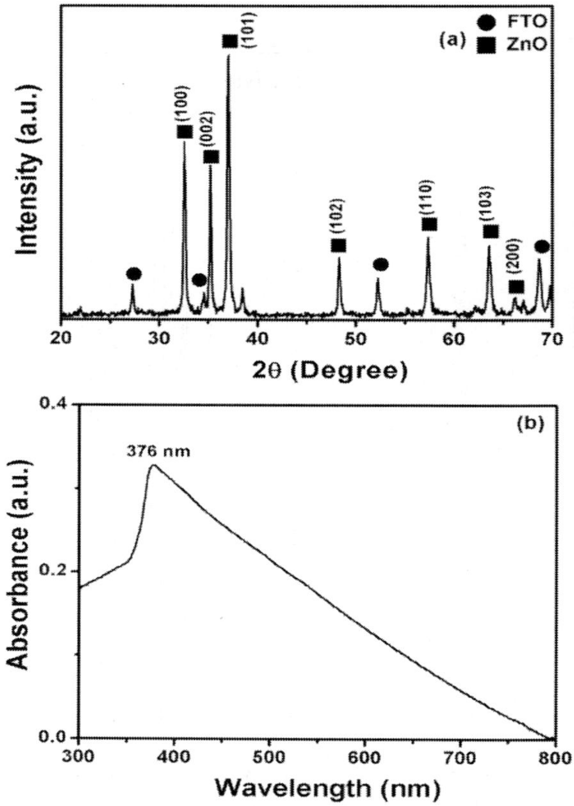

Figure 20: XRD patterns (a) and UV-Vis spectra (b) of nanospikes decorated ZnO sheets. Reprinted with permission from [Ameen S., 2012], Chem. Eng. J. 195 (2012) 307©2012, Elsevier Ltd.

The survey XPS spectrum (Fig. 21 (a)) of grown nanospikes decorated ZnO sheets shows the three strong binding energies of Zn $2p_{3/2}$, Zn $2p_{1/2}$ and O 1s along with small C 1s binding energy. The other binding energies peaks are not detected, indicating the presence of Zn and O without other impurities. However, the C1s binding energy at ~ 284.6 eV is usually used as calibration for other binding energies in the spectrum to avoid the specimen charging [103]. The Zn 2p spectrum of the doublet peaks with the binding energies of ~1021 eV and ~1045 eV are shown Fig 21(b),corresponding to Zn $2p_{3/2}$ and Zn $2p_{1/2}$ in better symmetry, respectively. These binding energies and the difference between two binding energies to ~24 eV are attributed to the typical lattice Zinc in ZnO [104]. The peak at ~1021 eV is associated with the Zn^{2+} in ZnO wurzite structure [105]. Moreover, Zn 2p binding energy and the binding energy difference values confirm that Zn atoms are in +2 oxidation state in ZnO. The deconvolution of O 1s XPS spectrum (Fig. 21(c)) exhibits the main peak at ~528.3 eV along with three resolved peaks at ~529.2 eV, ~530.1 eV and ~531.1 eV. The higher and lower binding energy component at ~528.3 eV and ~529.2 eV are attributed to O_2^- ions on the wurtzite structure of the hexagonal Zn^{2+} ions [106]. Every O_2^- ions are surrounded by Zn atoms with the full appreciation of nearest neighbor O_2^- ions. The other binding energies at ~530.1 eV and ~531.1 eV are ascribed to few oxygen deficiency or oxygen vacancies within the ZnO materials. Therefore, highest binding energy of Zn 2p and O 1s spectra are associated with Zn^{2+} and O_2^- ions which form Zn-O bonds in ZnO crystals.

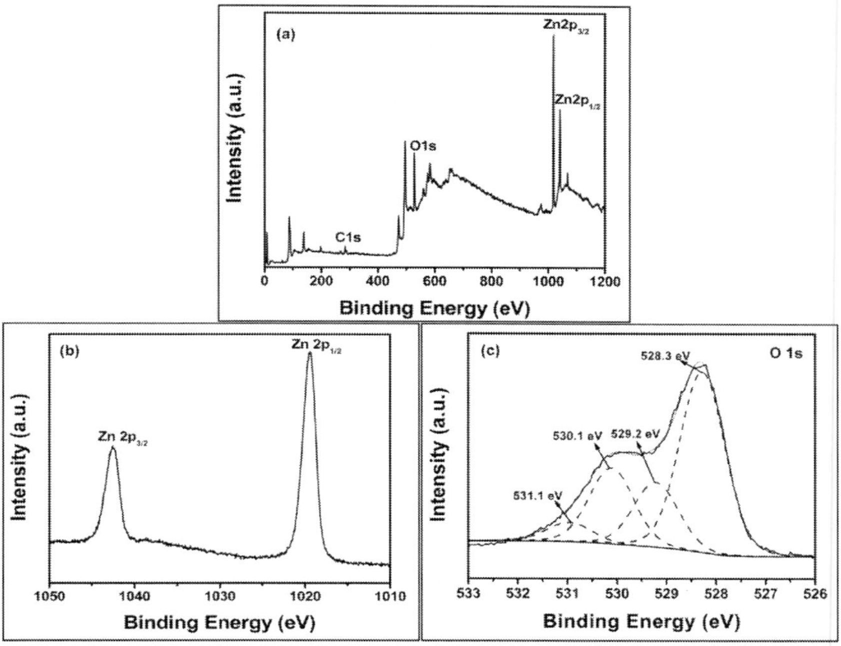

Figure 21: Survey (a), Zn 2p (b) and O 1s (c) XPS spectra of nanospikes decorated ZnO sheets. Reprinted with permission from [Ameen S., 2012], Chem. Eng. J. 195 (2012) 307 ©2012, Elsevier Ltd.

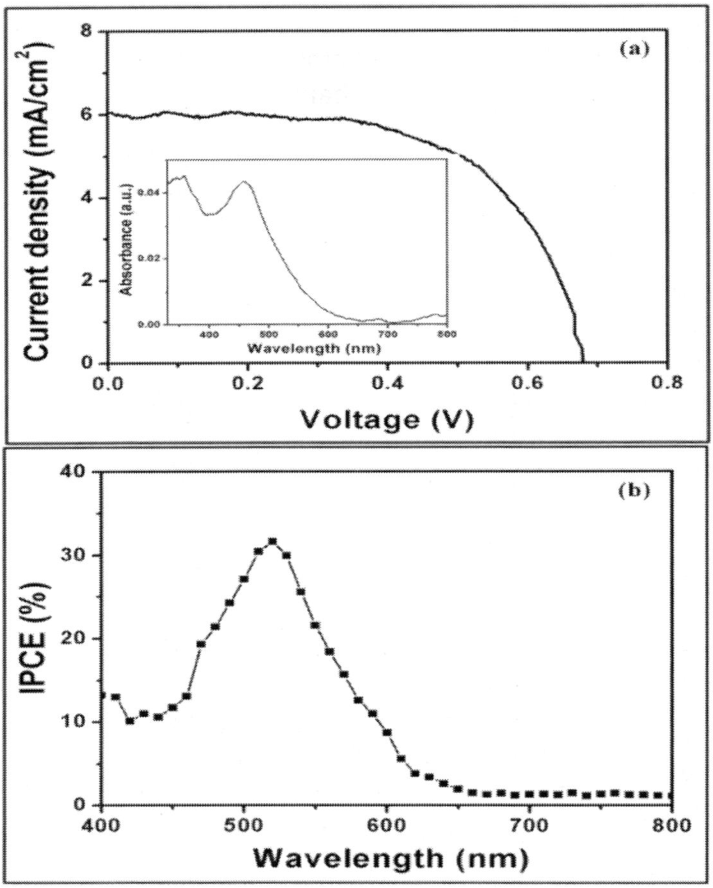

Figure 22: J-V curve (a) and IPCE (b) curve of the DSSC fabricated with nanospikes decorated ZnO sheets photoanode. Inset (a) shows the UV-Vis spectrum of desorbed dye from the nanospikes decorated ZnO sheets photoanode. Reprinted with permission from [Ameen S., 2012], Chem. Eng. J. 195 (2012) 307 ©2012, Elsevier Ltd.

The J-V curve (Fig. 22(a)) of DSSC fabricated with nanospikes decorated ZnO sheets photoanodes is demonstrated under the light intensity of 100 mW/cm^2 (1.5 AM). The fabricated DSSC with the photoanode of nanospikes decorated ZnO sheets has achieved the overall conversion efficiency of ~2.51% with the reasonably high J_{SC} of ~6.07 mA/cm^2, V_{OC} of ~0.68 V and FF of ~0.60. The relatively high J_{SC} is associated to high dye absorption through nanospikes decorated ZnO sheets morphology, resulting from the high amount of dye absorption (2.05×10^{-7} mol/cm^2) which is calculated by area integration of the maximum absorbance in the UV-Vis spectrum of desorbed dye from the photoanode (as shown in inset of Fig. 22(a)). The unique morphology of the prepared nanospikes decorated ZnO sheets might pronounce the charge collection and transfer properties of electrode due to the presence of standing spikes on the ZnO sheets [107]. The improved V_{OC} and FF of DSSC might attribute to the reduced charge recombination and the series resistance by the photoanode of nanospikes decorated ZnO sheets. As compared to the reported DSSCs based on ZnO nanostructures photoanodes, the nanospikes decorated ZnO sheets photoanode based DSSC shows the significantly higher conversion efficiency and J_{SC} [108]. It has been estimated that the conversion efficiency and J_{SC} are enhanced by ~40% and ~25% as compared to reported values. In this case, the sheets morphology of ZnO display highly uniform and the standing nanospikes might considerably facilitates the electrons transfer at the interface of the conduction and the electrolyte layer. The fabricated DSSC with photoanode of nanospikes decorated ZnO sheets attains the moderate IPCE of ~31.8%, as shown in Fig. 22 (b), which is probably originated from the larger amount of dye-loading through large surface area of sheet and the standing spikes of photoanode. Moreover, the presence of nanospikes on ZnO sheets might efficiently enhance the electron transport and reduces the recombination rate to high IPCE and J_{SC} value 109].

DOPING OF ZNO FOR IMPROVED ELECTRICAL AND PHOTOVOLTAIC PROPERTIES

One of the modifications is still in the developing stage called doping of ZnO nanomaterials by metals like F, Cu, Ag, Ga, Al, In, Sn and Sb which usually tailors the chemical, conductive and the electrical properties of ZnO nanomaterials [110]. The metal doping is an effective procedure to modify the grain size, orientation and the conductivity and could greatly influence the crystalline, optical and the electrical properties of the ZnO nanostructures. Among various metal doping, Sn-ion is recently known as promising dopant to ZnO nanomaterials for enhancing the electrical and optical properties [111]. Tsay et al. [112] prepared the Sn doped ZnO thin films coated glass substrates and investigated the effects of Sn doping on the crystallinity, microstructures and the optical properties of ZnO thin film. Several reports are available on the preparation of the Sn doped ZnO thin films and the effects of Sn doping on grain size, vibrational structure, optical and the structural properties of ZnO thin film substrates [113]. Ameen et al recently reported the doping of ZnO nanostructures with Sn ion by simple hydrothermal process for the fabrication of DSSC [114].

Sn Doped Zno Nanostructures for Solar Cell Performance

The synthesized ZnO and Sn-ZnO nanostructures are morphologically characterized by the FESEM images, as shown in Fig. 23. The ZnO nanostructure shows the irregular, non-uniform and highly aggregated nanoparticles with the average size of range ~150–200 nm. After Sn-ion doping, the ZnO nanostructures have dramatically arranged into the spindle shaped morphology and each Sn-ZnO spindle comprises of small aggregated nanoparticles with average size of 350 ±50 nm.

Figure 23: FESEM images of ZnO (a) and Sn-ZnO (b) nanostructures at low resolution, and FESEM images of Sn-ZnO nanostructures at high resolution (c and d). Reprinted with permission from [Ameen S., 2012], Chem. Eng. J. 187 (2012) 351 ©2012, Elsevier Ltd.

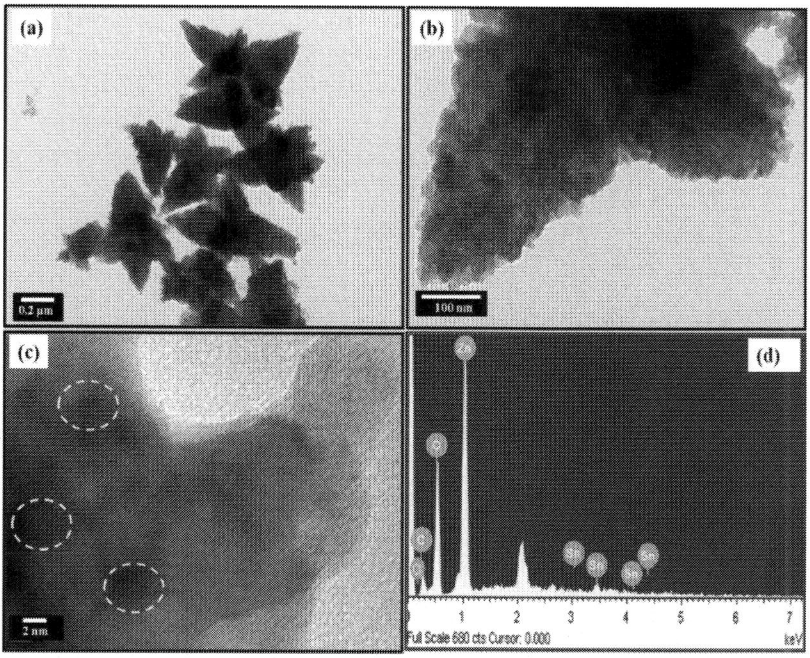

Figure 24: TEM images of Sn-ZnO nanostructures at low resolution (a and b), HRTEM image (c), and EDS spectrum of Sn-ZnO nanostructures (d). Reprinted with permission from [Ameen S., 2012], Chem. Eng. J. 187 (2012) 351 ©2012, Elsevier Ltd.

Fig. 24 (a, b) shows the high resolution TEM images of Sn-ZnO nanostructures which are completely consistent with FESEM observations. The aggregated ZnO nanoparticles arranged spindle shaped morphology is observed with some black spots or particles, clearly indicating the presence of the Sn-ions. The HR-TEM image of Sn-ZnO displays dark spots on the fringes which are expressed by the circles in Fig. 24(c). The morphological changes in Sn-ZnO nanostructures might due to the substantive influence of Sn-ion into ZnO nanostructures. The EDS spectrum (Fig. 24 (d)) obtains two high intense peaks (Zn & O) and single small peak (C) along with Sn peaks, again confirming the Sn-ion doping into the ZnO nanostructures.

From the UV-DRS spectra of ZnO and Sn-ZnO nanostructures (Fig. 25), the broad intense absorption edge from ~400 nm to lower wavelengths region is assigned to the charge-transfer process from the valence band to conduction band of ZnO [115]. After Sn-ion doping, the absorption wavelength of ZnO has significantly shifted from ~389 nm to ~406 nm and its band gap has changed from ~3.18 eV to ~3.05 eV. It has arisen due to the presence of interstitially embedded Sn-ion into ZnO nanomaterials. This small variation in band gaps again confirms the Sn-ion doping into ZnO nanomaterials.

The XPS survey (Fig. 26 (a)) spectrum displays all Zn 2p, O 1s and Sn 3d binding energy peaks with very small C 1s binding energy. Sn-ZnO nanomaterials show the doublet binding energies at ~1022.1 eV and ~1046.1 eV in Zn 2p XPS spectra (Fig. 26 (b)) which correspond to Zn $2p_{3/2}$ and Zn $2p_{1/2}$respectively. The energy difference between doublet binding energies is calculated to ~24 eV, which is in excellent agreement with the standard value of ~22.97 eV. The deconvoluted O 1s XPS presents the four binding energies peaks at ~533.6 eV, ~532.4 eV, ~530.8 eV and ~529.1 eV, as shown in Fig. 26 (c). The highest binding energy at ~533.6 eV is originated from the oxygen atoms chemisorbed at the surface of synthesized materials [116]. The binding energy at ~532.4 e V is ascribed to O_2^- ions (surface hydroxyl (OH) group) on the synthesized Sn-ZnO (in the oxygen deficient region) and the lowest binding energy at ~529.1 eV is attributed to O_2^- ions in the Zn-O structures The binding energy at ~530.8 eV is attributed to oxidized metal ions in the synthesized Sn-ZnO such as, O-Sn and O-Zn in the ZnO lattice. Sn 3d spectra (Fig. 26 (d)) presents the doublet binding energies at ~487.2 eV and ~496.7 eV, correspond to Sn $3d_{5/2}$ and Sn $3d_{3/2}$ respectively. The appearance of these peaks indicates the incorporation of Sn dopant in the form of O-Sn in the ZnO lattice [117], as deduced by O 1s XPS results. Moreover, the energy gap of ~9.5 eV is observed between these two peaks which resembles to the reported value [118]. It is observed that since no diffraction peaks corresponding to the SnO and SnO_2 are observed in the XRD spectra therefore, the O-Sn bonding could be considered as the substitutional doping of Sn-ions into the ZnO lattice.

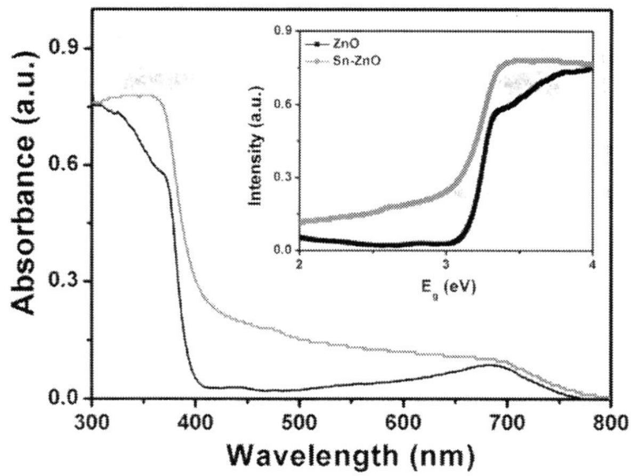

Figure 25: UV-DRS spectra of ZnO and Sn-ZnO nanostructures. Reprinted with permission from [Ameen S., 2012], Chem. Eng. J. 187 (2012) 351 ©2012, Elsevier Ltd.

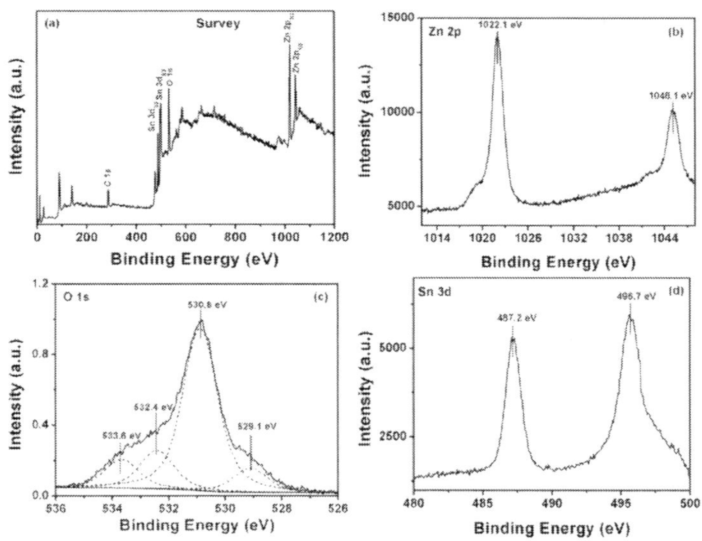

Figure 26: Survey (a), Zn 2p (b), O 1s (c), and Sn 3d (d) XPS spectra of Sn-ZnO nanostructures. Reprinted with permission from [Ameen S., 2012], Chem. Eng. J. 187 (2012) 351. ©2012, Elsevier Ltd.

DSSC fabricated (Fig. 27) with Sn-ZnO photoanode depicts reasonably high solar-to -electricity conversion efficiency of ~1.82% with J_{SC} of 5.1 mA/cm^2, V_{OC} of 0.786 V and FF of 0.45. While, DSSC with ZnO photoanode shows relatively low conversion efficiency of ~1.49% with J_{SC} of ~4.05 mA/cm^2, V_{OC} (~0.761 V) and FF of ~0.48. Noticeably, the conversion efficiency and J_{SC} are considerably enhanced by ~20% and ~21% respectively after Sn-ion doping into ZnO nanostructures. Moreover, it could be explained that Sn-ZnO nanostructures might due to the increase of high charge collection and the transfer of electrons at the interface of Sn-ZnO and the electrolyte layer. The dopants like Sn, is known to enhance the electrons transport capacity and electron mobility of ZnO nanomaterials [119]. Moreover, the Sn-ion doping into ZnO nanostructures might increase the specific surface area by lowering the particle size and arranging into spindle shaped morphology, which might contribute to high dye absorption. The increased photocurrent density and the improved photovoltaic performance might also result from high dye absorption and the improved electron transport by Sn-ZnO nanostructures, leading the enhancement in light harvesting efficiency and photo-excited electron transportation under sun light. Therefore, the arrangement of ZnO nanoparticles into Sn-ZnO spindle shaped, and good optical properties of Sn-ZnO are crucial to improve the conversion efficiency and the photocurrent density of the fabricated DSSCs.

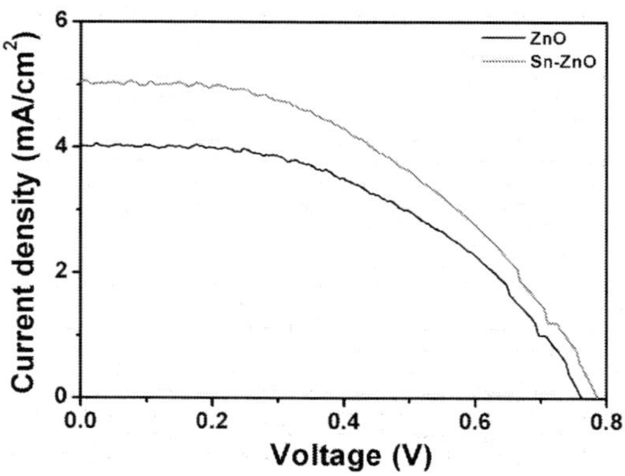

Figure 27: J-V curve of the DSSC fabricated with ZnO and Sn-ZnO nanostructures based photoanodes. Reprinted with permission from [Ameen S., 2012], Chem. Eng. J. 187 (2012) 351. ©2012, Elsevier Ltd.

Ga Doped Zno Nanostructures with Improved Electrical Properties

ZnO nanomaterials along with conjugated polymers like PANI, PPy and poly (3-alkylthiophene) comports the high quality organic/inorganic Schottky diodes [120]. Recently, ZnO-PANI films sandwiched between indium tin oxide (ITO) and a Pt electrode have displayed the linear I-V behavior [121]. The effects of Ga ion doping on ZnO NPs have been studied by Ameen et al on the basis of optical and electrical properties of the fabricated heterostructure devices [122]. The morphology of the synthesized ZnO and Ga-ZnO NPs were studied by FESEM and TEM analysis, as shown in Fig. 28 (a-d). The synthesized ZnO NPs obtain an average diameter of ~20–25 nm. After Ga ion doping, the average diameter increases to ~30–35 nm by the agglomeration of NPs due to entrapping and the substantive influence of Ga ion with ZnO NPs.

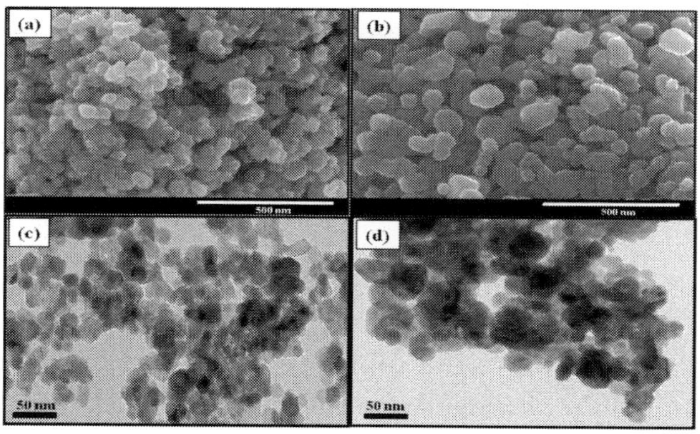

Figure 28: FESEM (a, b) and TEM (c, d) images of ZnO NPs and Ga-ZnO NPs. Reprinted with permission from [Ameen S., 2011], Microchim. Acta 172 (2011) 471. © 2012, Springerlink Ltd.

The optical properties of ZnO and Ga-doped ZnO NPs were studied by the UV-Vis spectra, as shown in Fig. 29 (a, b). ZnO and Ga-doped ZnO NPs present the absorption in the UV region with strong absorption peak at ~370 nm and ~378 nm respectively, corresponding to the characteristic band of wurtzite hexagonal ZnO nanomaterials [123]. A considerable red shift from ~370 nm to ~378 nm after Ga ion doping is seen in the absorption peak of Ga-ZnO NPs and results that the band gap of ZnO NPs has changed from ~3.4 eV to ~3.26 eV due to the presence of interstitially embedded Ga ion on the surface of ZnO NPs. Thus, the small variation in band gaps again confirms the Ga doping on the surface of ZnO NPs.

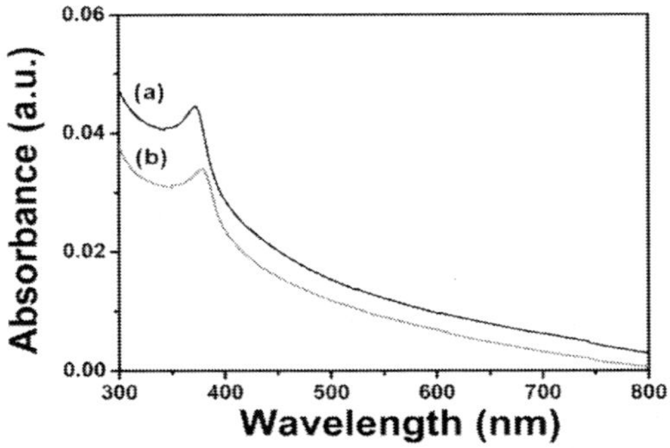

Figure 29: UV-Vis spectra of ZnO (a) and Ga-ZnO NPs (b). Reprinted with permission from [Ameen S., 2011], Microchim Acta 172 (2011) 471© 2012, Springerlink Ltd.

Fig. 30 shows the XPS of PANI/Ga-ZnO thin film electrodes. The Carbon (C 1s), oxygen (O 1s), nitrogen (N 1s) and Zinc (Zn 2p) peaks at ~284.4 eV, ~529.8 eV, ~400.9 eV and ~1019.4/1042.5 eV are taken to investigate the interaction between PANI and Ga-ZnO NPs. The deconvoluted C 1s peak at ~284.4 eV presents four resolved peaks at ~289.1, ~286.8, ~285.7 eV and ~284.9 eV (Fig. 30 (a)) and are ascribed to C = O/C–O bond, C–N^+/C = N^+ bond, neutral C–C/C–H bond of PANI backbone and C of the benzonoid ring showing a combination of protonation of imine and amine sites via shake-up processes [124]. Figure 30(b) shows the four O 1s XPS resolved peaks of PANI/Ga-ZnO thin film. The main peak at ~529.8 eV confirms the nature of oxygen atom originated from metal oxide [125] i.e. the oxide of ZnO NPs. Zn 2p XPS of PANI/Ga-ZnO thin film typically exhibits the doublet peaks at ~1019.4 eV/1042.5 eV (Fig. 30 d), suggests that Zn atoms are linked with oxide bond in Ga doped ZnO NPs. Moreover, Fig. 30(e) shows the Ga 2p peak at ~1116.2 eV which confirms the doping of ZnO with the Ga^{+2} oxidation state [126]. N 1s XPS spectrum of PANI/Ga-ZnO thin film (Fig. 30(c)) exhibits the bonding between imine group of PANI and Ga-ZnO. The centered peak at ~400.9 eV ascribes quinoid di-imine nitrogen of PANI [127]. The main peak at ~400.9 eV is resolved into four hide peaks at ~400.5, ~401.3, ~401.9 and ~402.8 eV which correspond to benzenoid di-amine nitrogen, quinoid di-imine nitrogen, positively charged nitrogen (−N⁺) and the protonated imine (=N⁺) respectively. It is known that the protonation of PANI produces electronic defects such as polarons or bipolarons which might form by the addition of protons to the neutral polymer chain. In this case, positively shifted binding energy at ~401.9 and ~402.8 eV might exhibit the participation of protonated N atom for the bond formation between PANI and Ga-ZnO. These two charged nitrogen species (−N⁺ and =N⁺) are originated from these defect states [128] and are observed in N 1s results of PANI/Ga-ZnO thin film. In conclusion, the PANI and Ga-ZnO are interacted to each other by the formation of partial hydrogen bonding between two charged nitrogen species (−N⁺ and =N⁺) of PANI and surface hydroxyl of Ga-ZnO.

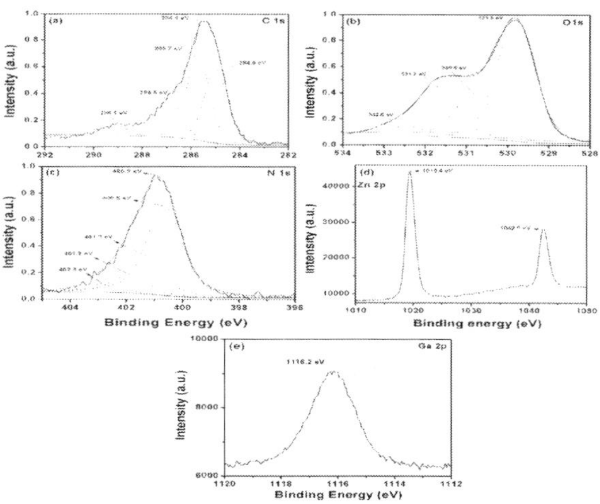

Figure 30: (a) C 1 s, (b) O 1 s, (c) N 1 s, (d) Zn 2p and (e) Ga 2p XPS spectra of PANI/Ga-ZnO thin film electrode. Reprinted with permission from [Ameen S., 2011], Microchim. Acta 172 (2011) 471©2012, Springerlink Ltd.

The I-V characteristics of Pt/PANI/ZnO and Pt/PANI/Ga-ZnO heterostructure devices are obtained at 298 K with applied voltage from −1 V to +1 V, shown in Fig. 31. Both the devices display non-linear and rectifying behavior of I–V curves due to the existence of Schottky barrier via a Schottky contact at the interfaces of Pt layer and

PANI-ZnO thin film layer. Pt/PANI/ZnO device shows almost Ohmic or very weak rectifying behavior (Fig. 31(a)) that attains very low turn-on voltage (~0.0005 V) with least current (~0.002 mA). Similarly, in forward bias, a breakdown voltage (~0.05 V) and high leakage current (~0.015 mA) indicate poor I-V characteristics of Pt/PANI/ZnO device. Whereas, Pt/PANI/Ga-ZnO device (Fig. 31 (b)) presents rectifying behavior of lower turn on voltage (~0.4 V) with least current (~0.09 mA) and breakdown voltage (~0.56 V) with high leakage current (~0.5 mA). The I-V properties of Pt/PANI/Ga-ZnO device are considerably better than the data reported elsewhere on PANI/ZnO and PANI based heterostructure devices [129]. Herein, the Ga ion doping to ZnO NPs might generate high density of minority charge carriers and the efficient charge movement at the junction of PANI and Ga-ZnO interfaces, resulting in the high leakage current with moderate turn on and breakdown voltage [130]. Additionally, the improved I-V properties might result from the molecular geometry of PANI chains and the increased electronic conduction by Ga ions in ZnO NPs which likely generate the hopping effect.

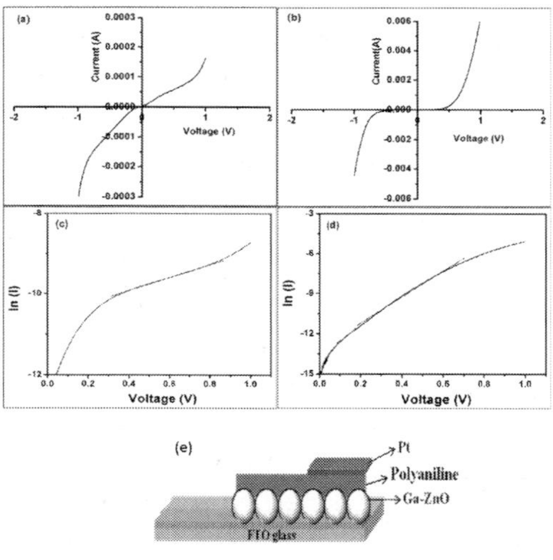

Figure 31: I-V characteristics of (a) Pt/PANI/ZnO (b) Pt/PANI/Ga- ZnO heterostructure device. ln(I) versus (V) plots of (c) Pt/PANI/ZnO and (d) Pt/PANI/Ga-ZnO heterostructure devices. (e) the schematic of Pt/PANI/Ga-ZnO heterostructure device. Reprinted with permission from [Ameen S., 2011], Microchim. Acta 172 (2011) 471©2012, Springerlink Ltd.

METAL OXIDES AS NANOFILLERS IN POLYMER ELECTROLYTES

The inorganic semiconductor especially metal oxides nanomaterials as nanofillers are conceived to improve the mechanical, thermal, interfacial, and ionic conductivity properties of the polymer electrolytes, which could effectively utilize for high performance solid-state DSSCs. In general, the introduction of inorganic NPs in the polymer alters the conduction mechanisms, which assigns to the ions conduction. In 1998, Croce et al. [131] studied the enhancement of the ionic conductivity of polymer electrolytes by the addition of TiO_2 and other NPs. Later, a ternary component polymer-gel electrolyte with TiO_2 NPs was prepared by Kang et al and explicated that these NPs led to a light-scattering effect [132]. The fabricated DSSC with the ternary component polymer electrolyte showed a high overall conversion efficiency of ~7.2% under 100 mW/cm^2. Falaras et al [133] developed the polymer composite electrolyte by the addition of commercial TiO_2 nanoparticles (NPs, P25, Degussa) consisted of polyethylene oxide (PEO), LiI, and I_2. The addition of TiO_2 NPs considerably prevented the re-crystallization and decreased the degree of crystallinity of PEO due to their large surface area. The differential scanning calorimetry (DSC) studies revealed that the introduction of TiO_2 NPs caused a significant increase in the glass transition temperature of PEO, which indicated the incorporation of the polymer to the inorganic oxide fillers. The fabricated DSSCs with TiO_2-PEO nanocomposite electrolyte achieved a reasonably high overall conversion efficiency of ~4.2% with a J_{SC} of ~7.2 mA/cm^2, V_{OC} ~0.664 V, and FF ~0.58 at ~65.6 mW/cm^2 [134]. Additionally, other research groups have also used TiO_2 NPs as nanofillers and explained the effects of nanofillers on different polymer electrolytes. Recently, Akhtar et al [135] reported a composite electrolyte of polyethylene glycol methyl ether (PEGME) and TiO_2 NPs and demonstrated the heat treatment effects on the properties of PEGME-TiO_2 composite electrolyte. It was found that the heat treatment was an essential step to improve morphology, amorphicity, and ionic conductivity of PEGME-TiO_2 composite electrolytes. From AFM images (Fig. 32), TiO_2 particles with ~20-30 nm size are well distributed on the PEGME matrix in the case of PEGME-TiO_2 composite film (Fig. 32(a). However, TiO_2 particles are aggregated and become a bigger size (40-

60 nm) on the polymer matrix in PEGME-TiO_2/80°C (Fig. 32 (b)). From the film roughness analysis (Fig. 32 (c, d)), it is observed that the surface roughness of the PEGME-TiO_2/80°C and PEGME-TiO_2 composites are estimated to be about ~23.1 nm and ~18.5 nm for the root mean square roughness (R_{rms}), and ~12 nm and 8~.8 nm for the average surface roughness (*Ra*), respectively. Generally, it has been well known that the low surface roughness of the polymer composite film is ascribed to the high-crystallized surface of thr composite materials. Therefore, the crystallinity of PEGME-TiO_2/80°C might lower than PEGME-TiO_2 because the former exhibits the higher R_{rms} and *Ra* value than later composite film. Consistently, the 3D AFM images (Fig. 32 (d)) of PEGME-TiO_2/80°C exhibit a highly rough surface morphology with non-uniformly distributed TiO_2 particles into the PEGME matrix, while a highly uniform and less rough surface is observed in PEGME-TiO_2 composite film before heating (Fig. 32 (c)). This rough morphology of PEGME-TiO_2/80°C might create free spaces and voids in which the I^-/I_3^- ions could easily migrate, which suggest the PEGME-TiO_2/80°C as excellent electrolyte materials. With the improved morphology of PEGME-TiO_2 composite, the electrolyte shows the high ionic conductivity of ~1.9 mS/cm as compared to the PEGME acid (1.2 mS/cm) and PEGME-TiO_2 (0.92 mS/cm) which results from the enhanced morphological properties in terms of high roughness and amorphicity after heating of PEGME-TiO_2. The Raman spectra (Fig. 33) of the PEGME-acid, PEGME-TiO_2, and PEGME-TiO_2/80°C composite electrolytes exhibit a significant peak at the range of 110-115 cm^{-1}, which is ascribed to the symmetric stretch of I_3^- species in redox electrolytes [136]. The heat treatment on PEGME-TiO_2 drastically increased the strong Raman peak, indicating a significant increase in the amount of I_3^- species in redox electrolytes. It might attribute to the increased bond strength between PEGME and TiO_2 and high roughness of the composite materials, which might help to absorb a large amount of the iodide couple as compared to PEGME-acid and PEGME-TiO_2 composite electrolytes. The increased intensity of peak suggests that a large amount of I_3^- species is formed in the PEGME-TiO_2/80°C composite electrolyte upon heat treatment. In general, the diffusional I^-/I_3^- ions migration in the redox electrolyte is responsible for the ionic conductivity of electrolyte, which causes electron exchange between ions by electronic conduction process [137]. The electronic conduction in redox electrolyte depends on the formation of I_3^- ions. Raman results

show the proportional relation between the ionic conductivity and concentration of I_3^- species of the composite electrolytes, which is directly related to relative intensity of Raman peak. The enhanced ionic conductivity of PEGME-TiO_2/80°C composite electrolyte might associate with the formation of high I_3^- species in redox electrolyte. However, low ionic conductivity in PEGME-acid and PEGME-TiO_2 composite electrolyte results from the low relative intensity of the Raman peak and less formation of I_3^- species in redox electrolyte. Therefore, a heat treatment step plays an essential role to prepare the improved composite electrolyte with enhanced ionic conductivity. DSSC fabricated with PEGME-TiO_2/80°C composite electrolyte shows the maximum overall conversion efficiency of ~3.1% with a J_{SC} of ~8.9 mA/cm^2, V_{OC} of ~0.625 volt, and FF of ~56.2%. The conversion efficiency and J_{SC} of DSSCs with PEGME-TiO_2/80°C composite electrolytes is higher than those of fabricated with PEGME-acid (~1.3%) and PEGME-TiO_2 (~2.4%) electrolytes. This could be expected from the enhanced ionic conductivity and enlargement of the amorphous phase of the polymer upon heat treatment. The heat treatment on PEGME-TiO_2 composites enhances the ionic conductivity and cross-linking properties of composite electrolyte, which are essential factors to achieve the high current density and high PV performance. Furthermore, Akhtar et al [138] investigated the effect of titania nanotubes (NTs) as nanofillers on the properties of PEG-based electrolytes and fabricated solid-state DSSCs. PEG-TiNT electrolytes with 10% of TiNTs exhibit the high penetration and complete filling into the pores of the TiO_2 film, as shown in Fig. 34. The XPS studies (Fig. 35) were carried out to elucidate the strong interaction between PEG and TiNTs. PEG-TiNT10 electrolyte shows the highest interaction between the titanium atoms of the NTs and the polymer network as compared to those of other PEG-TiNTs electrolytes. This results to the decrease in the crystallinity degree of the polymer after introduction of the NTs which achieves the highest ionic conductivity of ~2.4×10^{-3} S/cm. DSSC fabricated with PEG-TiNT composite electrolyte (Fig. 36) exhibits the maximum overall conversion efficiency of ~4.4% with $J_{SC}$of ~9.4 mA/cm^2, V_{OC} of ~0.73 V, and FF of ~0.65 under 100 mW/cm^2 irradiation. No significant decrease of the conversion efficiency for 30 days was observed in DSSCs fabricated with PEG-TiNT10 (inset of Fig. 36), indicating the high stability of the composite electrolytes. The lower current density in PEG-TiNT20 is due to its lower ion conductivity, lower penetration, and weak interaction between PEG to TiNTs. It is proved that the better penetration into the pores of the TiO_2layer was obtained at a ratio of TiNT and PEG in the composite electrolyte (PEG-TiNT10). Thus, due to the better interfacial contact between the electrolyte and TiO_2 layer, high ion conductivity is obtained, which enhances the photocurrent density. Moreover, the PEG-TiNT composite electrolytes might facilitate the movement of electrons in the redox (I^-/I_3^-) couple due to the fast electron transfer characteristics of NTs with less grain-boundary, in comparison to NPs. The other metal oxide nanomaterials, such as ZnO, SiO_2 and Al_2O_3, have been introduced as nanofillers to different polymer electrolytes [139]. Caruso et al. prepared the polymer composite electrolyte with the composition of PEO-poly (vinylidenefluoride) (PVDF) and SiO_2 NPs and applied as a solid electrolyte for solid-state DSSCs [140]. These kinds of solid state electrolytes presented high viscosity to the solid state electrolyte with TiO_2-based polymer composites. They used a vacuum technique for introducing the composite polymer electrolytes into the dye-sensitized TiO_2 electrode which showed that the vacuum method exhibited a better performance than those prepared via the conventional drop casting method. This approach improved the fulfilling of the photoelectrode with a solid electrolyte by vacuum technique, but the optimization of the electrolyte composition is still an important issue. In this regards, Xia et al utilized ZnO NPs as nanofillers for preparing the composite polymeric electrolyte of poly (ethylene glycol methyl ether) (PEGME) [141]. The PEGME was first grafted onto the surface of ZnO NPs through covalent bond formation by a chemical process. The solid composite electrolyte consisted of KI and I_2 dissolved in PEGME and ~24 wt% of the polymer-grafted ZnO NPs. The obtained prepared electrolyte showed that the ionic conductivity increased as the salt concentration increased and reached a maximum value of $\sim 3.3\times10^{-4}$ S/cm and then decreased, acting as a classical polymer electrolyte system. DSSC fabricated with polymer-grafted NPs electrolyte presented the lower conversion efficiency of ~3.1% compared to that of DSSC with a liquid electrolyte (~4.0%). After the addition of polymer-grafted ZnO NPs in liquid electrolyte, the V_{OC} of DSSC increased by ~0.13 V while the J_{SC} decreased, this was probably due to the high viscosity of the gel electrolyte. Another report addressed the new polymer electrolyte system of Al_2O_3 NPs with different sizes and a PVDF derivative and polyacrylonitrile in an ionic-liquid-based electrolyte

[142]. The diffusion coefficient of I_3^- ions altered by the addition Al_2O_3 NPs. The variation in sizes of Al_2O_3 NPs greatly influenced the charge transfer rate at the electrolyte and semiconducting layer interfaces. In this report, the imidazolium cations might adsorb on the NP surface, which might help in the charge transfer at counter and anions I^-/I_3^- gather around them. Some researchers have recently used clay-like NPs as nanofillers in the polymer electrolytes and applied to the DSSCs [143]. Nogueira et al. [144] examined the incorporation of a montmorillonite (MMT) derivative to a polymer electrolyte based on a poly-(ethylene oxide) copolymer, the plasticizer GBL, and Li I/I_2. The improved ionic conductivity of the composite electrolyte attributed to the large number of charge carriers introduced into the complex after the addition of the clay. The addition of 5 wt% MMT promoted the increase in the mechanical stability of the nanocomposite polymer electrolyte film, resulted in the lower deformation as compared to the film without any clay. From their observations, it was found that the addition of MMT clay to the plasticized polymer electrolyte not only increased the ionic conductivity but also improved the solidification of the electrolyte. These improvements led to the mechanical stability of the polymer composite films and the stability of DSSCs as well. DSSCs fabricated with the nanocomposites polymer electrolyte showed reasonable conversion efficiencies of ~1.6% and ~3.2% at 100 mW/$cm^2$and 10 mW/cm^2, respectively. The device presented very poor FF values of ~0.40 at 100 mW/cm^2, which was attributed to the low penetration of the composite electrolyte into the pores of the TiO_2film. The MMT clay as nanofillers was also used by Lin et al. They prepared the nanocomposites of poly (nisopropylacrylamide) with MMT clay to a liquid electrolyte system as a gelator and applied as solid polymer electrolyte. The poly (nisopropylacrylamide)-MMT electrolyte-based DSSC achieved a relatively high conversion efficiency of ~5.4% with a J_{SC} of ~12.6 mA/cm^2, V_{OC} of ~0.73 V, and FF of ~0.59, whereas the DSSC prepared with the electrolyte gelled with the pure polymer presented lower photovoltaic parameters of J_{SC} (~7.28 mA/cm^2), V_{OC} (~0.72 V), FF (0.60), and conversion efficiency (~3.2%) at 100 mW/cm^2. From the electrochemical impedance spectroscopy, a considerable decrease in impedance values was observed by DSSC fabricated with nanocomposite-gelled electrolyte. The impedance at the electrolyte/dye-coated TiO_2 interface, and the Nernstian diffusion within the electrolytes were decreased, resulted in the high photocurrent density

leading to the high performance of DSSCs. They also investigated the molar conductivity of the nanocomposite-gelled electrolytes to explain the high ionic conductivity and improved electrochemical behavior of electrolyte.

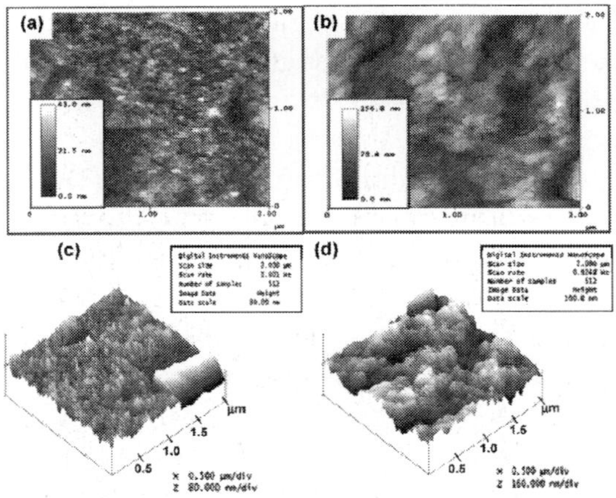

Figure 32: Topographic and three-dimensional AFM images of (a), (c) PEGME-TiO_2 and (b), (d) PEGMETiO_2/80C. Reprinted with permission from [Akhtar, 2011], Mater. Chem. Phys. 127 (2011) 479.© 2011, Elsevier Ltd.

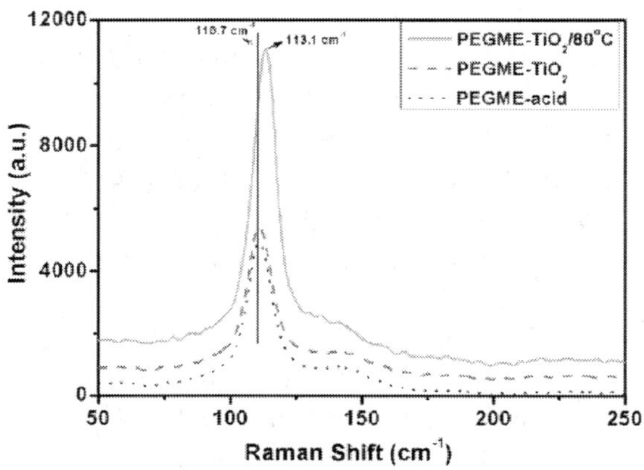

Figure 33: Raman spectra of PEGME-acid, PEGME–TiO_2 and PEGME–TiO_2/80 ∘C composite electrolytes. Reprinted with permission from [Akhtar., 2011], Mater. Chem. Phys. 127 (2011) 479. © 2011, Elsevier Ltd.

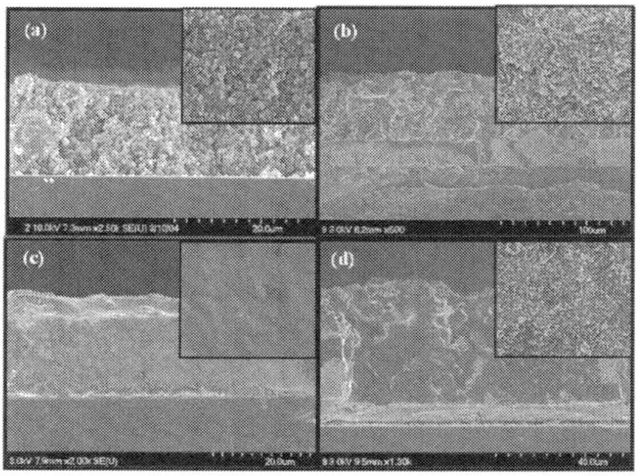

Figure 34: Cross-section and top view (inset) FE-SEM images of the TiO2 thin film (a) before electrolyte filling and after introducing composite electrolytes of (b) PEG-TiNT5, (c) PEG-TiNT10, and (d) PEG-TiNT20. Reprinted with permission from [Akhtar., 2007], Electrochem. Commun. 9 (2007) 2833.© 2007, Elsevier Ltd.

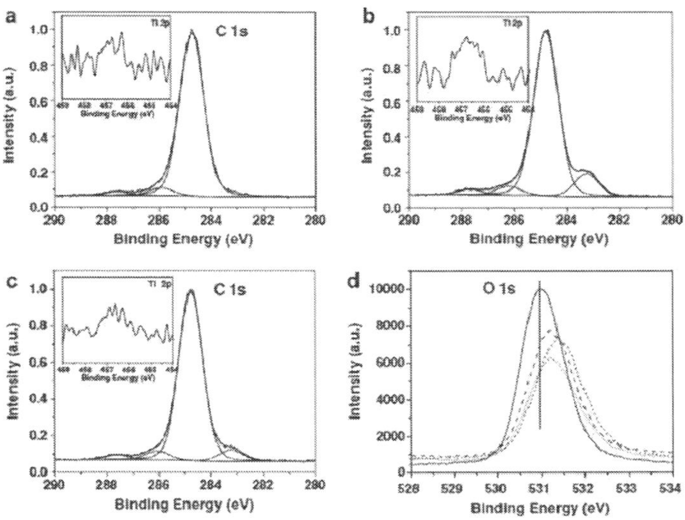

Figure 35: XPS spectra of the composite electrolytes. (a) PEG-TiNT5, (b) PEG-TiNT10, (c) PEG-TiNT20, and (d) O 1 s of PEG (—), PEG-TiNT5 (- - - -) PEG-TiNT10 (.....) and PEG-TiNT20 (-.-.-). Reprinted with permission from [Akhtar., 2007], Electrochem. Commun. 9 (2007) 2833.© 2007, Elsevier Ltd.

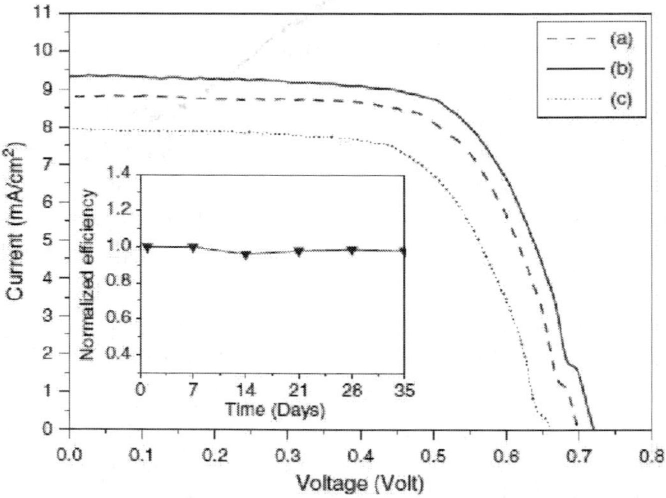

Figure 36: Current–voltage characteristics of DSSC fabricated with composite electrolytes of (a) PEG-TiNT5, (b) PEG-TiNT10, and (c) PEG-TiNT20. Inset shows the stability test of DSSC fabricated with composite electrolyte of PEG-TiNT10. Reprinted with permission from [Akhtar, 2007], Electrochem. Commun. 9 (2007) 2833.© 2007, Elsevier Ltd.

CONCLUSIONS

In summary, the morphological, structural, crystalline, optical, electrical and photovoltaic properties of conducting polymers, nanocomposites of conducting polymer/inorganic nanomaterials and semiconducting metal oxides have been discussed. The PANI nanocomposites with semiconducting materials have shown the improved penetration and optoelectronic properties, and applied for the electrical and electronic application such as diodes and solar cells. Here, the uniform distribution of CdS nanomaterials effectively improves the electronic state of PANI like polarons and bipolarons for the high charge carriers and enhances the charge transfer. The unique conducting polymers, particularly PANI nanomaterials have been used as hole transporting material and as counter electrodes for the applications of DSSCs. The metal oxide semiconducting nanomaterials, particularly TiO_2 and

ZnO nanomaterials, in terms of morphology, surface properties, dye absorption and application in DSSCs are extensively summarized. Various morphologies of metal oxides nanostructures greatly affect the performances of dye absorption, electrical, electrochemical, and photovoltaic devices. The metal oxides semiconducting nanomaterials with different morphologies and sizes enhance the surface-to-volume ratio and produce the highly advanced photoanodes for the efficient DSSCs. The morphologies of metal oxides semiconducting considerably influence the dye absorption, light harvesting and results in increased electron transfer and reduce the recombination rate during the operation of DSSCs. The photovoltaic properties such as J_{SC}, V_{OC}, FF, and conversion efficiency have significantly improved by altering the sizes and shapes of the metal oxides semiconductors. The chapter also summarizes the use of various metal oxide semiconducting nanomaterials as nanofillers in polymer electrolytes and describes their effect on the properties of polymer electrolytes and the performances of DSSCs. The introduction of metal oxide semiconducting nanomaterials into the polymer matrix has significantly improved the amorphicity, mechanical, thermal and ionic conductivity of polymer electrolytes. The chapter includes some of the polymer composite electrolytes and their photovoltaic properties for DSSCs.

REFERENCES

1. Shirakawa H. The discovery of polyacetylene film: the dawning of an era of conducting polymers. Angewandte Chemie International Edition 2001;40:2575–2580.
2. MacDiarmid AG. Synthetic metals: a novel role for organic polymers. Angewandte Chemie International Edition 2001; 40:2581–2590.
3. Gerard M, Chaubey A, Malhotra BD. Biosens. Application of conducting polymers to biosensors. Biosensors and Bioelectronics 2002;17:345-359.
4. Friend RH, Gymer RW, Holmes AB, Burroughes JH, Marks RN, Taliani C, Bradley DD C, Santos DA, Brédas JL, Lögdlund M, Salaneck WR. Electroluminescence in conjugated polymers. Nature 1999; 397:121-128.

5. Sirringhaus H, Tessler N, Friend RH. Integrated Optoelectronic Devices based on. Conjugated Polymers. Science 1998; 280:1741-1744.

6. Morrin A, Wilbeer F, Ngamna O, Moulton SE, Killard AJ, Wallace GG, Novel biosensor fabrication methodology based on processable conducting polyaniline nanoparticles. Electrochemistry Communication 2005; 7:317-322.

7. Jang J, Bae J, Lee K. Synthesis and characterization of polyaniline nanorods as curing agent and nanofiller for epoxy matrix composite, Polymer 2005;46:3677-3684.

8. Plesu N, Ilia G, Pascariu A, Vlase G. Preparation, degradation of polyaniline doped with organic phosphorus acids and corrosion essays of poly aniline-acrylic blends. Synthetic Metals 2006; 156:230-238.

9. Han DH, Lee HJ, Park SM. Electrochemistry of conductive polymers XXXV: Electrical and morphological characteristics of polypyrrole films prepared in aqueous media studied by current sensing atomic force microscopy. Electrochimica Acta 2005;50:3085-3092.

10. Johanson U, Marandi A, Tamm T, Tamm J. Comparative study of the behavior of anions in polypyrrole films. Electrochimica Acta 2005;50:1523-1528.

11. Krivan E, Peintler G, Visy C. Matrix rank analysis of spectral studies on the electropolymerisation and discharge process of conducting polypyrrole/dodecyl sulfate films. Electrochimica Acta 2005;50:1529-1535.

12. Chehimi MM, Abel ML, Perruchot C, Delamar M, Lascelles SF, Armes SP. The determination of the surface energy of conducting polymers by inverse gas chromatography at infinite dilution. Synthetic Metals 1999;104:51-59.

13. Khomenko VG, Barsukov VZ, Katashinskii AS. The catalytic activity of conducting polymers toward oxygen reduction. Electrochimica Acta 2005;50:1675-1683.

14. Hien NTL, Garcia B, Pailleret A, Deslouis C. Role of doping ions in the corrosion protection of iron by polypyrrole films. Electrochimica Acta 2005;50:1747-1755.

15. Liu L, Zhao Y, Zhou Q, Xu H, Zhao C, Jiang Z. Nano-polypyrrole supercapacitor arrays prepared by layer-by-layer assembling method in anodic aluminum oxide templates. Journal of Solid State Electrochemistry 2007;11:32-37.
16. Krivoshei IV, Skorobogatov VM. "Polyacetylene and Polyarylenes; Polymer Monographs." Vol. 10,Gordon and Breach Science Publishers, Philadelphia, PA, 1991
17. Kovacic P, Jones MB. Dehydro coupling of aromatic nuclei by catalyst-oxidant systems: poly(p-phenylene). Chemical Reviews 1987;87:357-379.
18. Shi G, Xue G, Li C, Jin S, Yu B. Uniaxial oriented poly(p-phenylene) fibrils and films. Macromolecules 1994;27:3678-3679.
19. Roth S, Bleier H. Solitons in polyacetylene. Advances in Physics1987;36:385-462.
20. Jacky JWY, Tang BZ. Functional polyacetylenes. Account of Chemical Research 2005;38:745-754.
21. Asato AE, Liu RSH, Rao VP, Cai YM. Azulene-Containing Donor-Acceptor Compounds as Second-Order Nonlinear Chromophores. Tetrahedron Letters 1996;37:419-422.
22. Taudi H, Bernede JC, Valle D, Bonnet MAA, Morsli M. Influence of the electrochemical conditions on the properties of polymerized carbazole. Journal of Materials Science 2001;36:631-634.
23. Baba A, Onishi K, Knoll W, Advincula RC. Investigating Work Function Tunable Hole-Injection/Transport Layers of Electrodeposited Polycarbazole Network Thin Films. Journal of Physical Chemistry B 2004;108:18949-18955.
24. Pud AA. Stability and degradation of conducting polymers in electrochemical systems. Synthetic Metals 1994;66(1):1–18.
25. Min G. Conducting polymers and their applications in the film industry: polyaniline polyimide blended films. Synthetic Metals 1999;102(1–3):1163–1166.
26. Davies JE, Less RJ, May I, Rawson JM. Isolation of the first diselenadiazolyl complex, Pd3[PhCNSeSeN]2[PPh3]4 2PhMe New J. Chem. 1998;22:763-765
27. Hahiwara T, Demura T, Iwata K. Synthesis and properties of electrically conducting polymers from aromatic amines. Synthetic Metals 1987;18:317-322.

28. Snauwaert P, Lazzaroni R, Riga J, Verbist JJ. Electronic structure of polyaniline and substituted derivatives. Synthetic Metals 1987; 18:335-340.

29. Cho MS, Cho YH, Choi HJ, Jhon MS. Synthesis and electrorheological characteristics of polyaniline-coated poly(methyl methacrylate) microsphere: Size effect. Langmuir 2003;19:5875-5881.

30. Jindal Z, Verma NK. Electrochemical template–assisted fabrication of CdS micro/nanostructures. Physica E 2009;41:1752–1756.

31. Soudi R, Kamal M, Shabaka AA, Abdelrazek EM, Eisa W. Synthesis, characterization and spectroscopic studies of CdS/polyaniline core/shell nanocomposites. Synthetic Metals 2010;160:479–484.

32. Xi Y, Zhou J, Guo H, Cai C, Lin Z. Enhanced photoluminescence in core-sheath CdS–PANI coaxial nanocables: a charge transfer mechanism. Chemical Physics Letters 2005;412:60-63.

33. Seoudi R, Shabaka AA., Kamal M, Abdelrazek EM, Eisa H. Dependence of structural, vibrational spectroscopy and optical properties on the particle sizes of CdS/polyaniline core/shell nanocomposites. Journal of Molecular Structure 2012;10(13):156-162.

34. Raut BT, Chougule MA, Sen S, Pawar RC, Lee CS, Patil VB. Novel method of fabrication of polyaniline–CdS nanocomposites: Structural, morphological and optoelectronic properties. Ceramics International 2012;38:3999-4007.

35. Ameen S, Akhtar MS, Kim YS, Shin HS. Synthesis and electrochemical impedance properties of CdS nanoparticles decorated polyaniline nanorods. Chemical Engineering Journal 2012;181–182:806–812.

36. Boyer MI, Quillard S, Louarn G, Froyer G, Lefrant S. Vibrational study of the FeCl3-doped dimer of polyaniline; a good model compound of emeraldine salt. Journal of Physical Chemistry B 2000;104:8952–896.

37. Mazeikiene R, Statino A, Kuodis Z, Niaura G, Malinauskas A. In situ Raman spectroelectrochemical study of self-doped polyaniline degradation kinetics. Electrochemistry Communications 2006;8:1082–1086.

38. Khiew PS. Huang NM, Radiman S, Ahmad MS. Synthesis and characterization of conducting polyaniline-coated cadmium sulphide nanocomposites in reverse microemulsion. Materials Letters 2004;58:516–521.

39. Shimano JY, MacDiarmid AG. Polyaniline, a dynamic block copolymer: key to attaining its intrinsic conductivity. Synthetic Metals 2001;123:251–262.

40. Monkman AP, Stevens GC, Bloor D. X-ray photoelectron spectroscopic investigations of the chain structure and doping mechanisms in polyaniline. Journal of Physics D: Applied Physics 1991;24:738–743.

41. Wu MS, Wen TC, Gopalan A. Electrochemical copolymerization of diphenylamine and anthranilic acid with various feed ratios. Journal of the Electrochemical Society 2001;148:D65–D73.

42. Ameen S, Akhtar MS., Kim YS, Yang OB, Shin HS. Electrical and structural characterization of plasma polymerized polyaniline/TiO2 heterostructure diode: a comparative study of single and bilayer TiO2 thin film electrode. Journal of Nanoscience and Nanotechnology 2011;11:3306–3313.

43. Glidle A, Swann MJ, Hadyoon CS, Cui L, Davis J, Ryder KS, Cooper JM. XPS assaying of electrodeposited copolymer composition to optimise sensor materials. Journal of Electron Spectroscopy and Related Phenomena 121 (2001) 131–148.

44. Suzer S, Birer O, Sevil UA, Guven O. XPS investigations on conducting polymers. Turkish Journal of Chemistry 1998;22:59–65.

45. Duran JDG, Guindo MC, Delgado AV, Caballero FG. Surface chemical analysis and electrokinetic properties of synthetic spherical mixed zinc–cadmium sulfides, Journal of Colloids and Interface Science 1997;193:223–233.

46. Fiordiponti P, Pistoia G. An impedance study of polyaniline films in aqueous and organic solutions. Electrochimica Acta 1989;34:215–221. (b) Vorotyntsev MA, Badiali JP, Inzelt G. Electrochemical impedance spectroscopy of thin films with two mobile charge carriers: effects of the interfacial charging. Journal of Electroanalytical Chemistry 1999;472:7–19.

47. Hagfeldt A, Gratzel M. Light-Induced Redox Reactions in Nanocrystalline Systems. Chemical Reviews 1995;95:49-68.
48. Cheng YJ, Yang SH, Hsu CS. Synthesis of conjugated polymers for organic solar cell applications, Chemical Reviews 2009;109:5868–5923.
49. Jaglarz J, Wagner T, Cisowski J, Sanetra J. Ellipsometric studies of carbazole-containing polymer layers. Optical Materials, 2007;29:908-912.
50. Kudo N, ShimazakiY, Ohkita H, Ohoka M, Ito S. Organicinorganic hybrid solar cells based on conducting polymer and SnO2 nanoparticles chemically modified with a fullerene derivative. Solar Energy Materials Solar Cells 2007;91:1243-1247.
51. Woo S, Jeong JH, Lyu HK, Jeong S, Sim JH, Kim WH, Han YS, Kim Y. Hybrid solar cells with conducting polymers and vertically aligned silicon nanowire arrays: The effect of silicon conductivity. Physica B: Condensed Matter 2012;407:3059-3062.
52. Tan F, Qu S, Wu J, Wang Z, Jin L, Bi Y, Cao J, Liu K, Zhang J. Wang Z., Electrodepostied polyaniline films decorated with nano-islands: Characterization and application as anode buffer layers in solar cells. Solar Energy Materials Solar Cells 2011;95:440-445.
53. Chang MY, Wu CS, Chen YF, Hsieh BZ, Huang WY, Ho KS, Hsieh TH, Han YK. Polymer solar cells incorporating one-dimensional polyaniline nanotubes. Organic Electronics 2008; 9:1136-1139.
54. Lim TH, Oh KW, Kim SH. Self-assembly supramolecules to enhance electrical conductivity of polyaniline for a flexible organic solar cells anode. Solar Energy Materials Solar Cells 2012;101:232-240.
55. Bejbouji H, Vignau LL, Miane J, Dang MT, Oualim EM, Harmouchi M, Mouhsen A. Polyaniline as a hole injection layer on organic photovoltaic cells. Solar Energy Materials Solar Cells. 2010;94:176-181.
56. Zhu S., Wei W., Chen X., Jiang M., Zhou Z. Hybrid structure of polyaniline/ZnO nanograss and its application in dye-sensitized solar cell with performance improvement. Journal of Solid State Chemistry 2012;190:174-179.

57. Ameen S, Akhtar MS, Song M, Kim YS, Shin HS. Plasma Deposited Polyaniline on Ruthenium Dye Sensitized Zno Thin Film for Dye Sensitized Solar Cell. Advanced Science Letters 2012: doi:10.1166/asl.2012.3305.
58. Malherbe R, Martinez J, Reguera E, Navarro E. Journal of Materials Science 1992;28:274.
59. Ameen S, Akhtar MS, Kim G-S, Kim YS, Yang O-B, Shin HS. Plasma-enhanced polymerized aniline/TiO2 dye-sensitized solar cells. Journal of Alloys and Compounds 2009;487:382–386.
60. Bisquert J. Theory of the Impedance of Electron Diffusion and Recombination in a Thin Layer. Journal of Physical Chemistry B 2002;106:325-333.
61. J.R. Macdonald, Impedance Spectroscopy, John Wiley & Sons, New York, 1987.
62. Brad AJ, Faulkner LR. Electrochemical Methods: Fundamentals and Applications, John Wiley & Sons, New York, 1980, p. 350.
63. Yeh MH, Lee CP, Chou CY, Lin LY, Wei HY, Chu CW, Vittal R, Ho KC. Conducting polymer-based counter electrode for a quantum-dot-sensitized solar cell (QDSSC) with a polysulfide electrolyte. Electrochimica Acta, 2011;57:277-284.
64.] Lee KM, Chen PY, Hsu CY, Huang JH, Ho WH, Chen HC, Ho KC. A high-performance counter electrode based on poly(3,4-alkylenedioxythiophene) for dye-sensitized solar cells. Journal of Power Sources. 2009;188:313-318.
65. Lee KM, Chiu WH, Wei HY, Hu CW, Suryanarayanan V, Hsieh WF, Ho KC. Effects of mesoscopic poly(3,4-ethylenedioxythiophene) films as counter electrodes for dye-sensitized solar cells. Thin Solid Films, 2010;518:1716-1721
66. Maiaugree W, Pimanpang S, Towannang M, Saekow S, Jarernboon W, Amornkitbamrung V. Optimization of TiO2 nanoparticle mixed PEDOT–PSS counter electrodes for high efficiency dye sensitized solar cell. Journal of Non-Crystalline Solids 2012; In Press: doi:10.1016/j.jnoncrysol.2011.12.104.
67. Chen J, Li B, Zheng J, Zhao J, Jing H, Zhu Z. Polyaniline nanofiber/carbon film as flexible counter electrodes in platinum-free dye-sensitized solar cells. Electrochimica Acta 2011; 56:4624-4630.

68. Li Q, Wu J, Tang Q, Lan Z, Li P, Lin J, Fan L. Application of microporous polyaniline counter electrode for dye-sensitized solar cells. Electrochemistry Communications 2008; 10:1299-1302.

69. Zhang J, Hreid T, Li X, Guo W, Wang L, Shi X, Su H, Yuan Z. Nanostructured polyaniline counter electrode for dye-sensitised solar cells: Fabrication and investigation of its electrochemical formation mechanism. Electrochimica Acta, 2010;55:3664-3668.

70. Wang G, Xing W, Zhuo S. The production of polyaniline/graphene hybrids for use as a counter electrode in dye-sensitized solar cells. Electrochimica Acta 2012;66:151-157.

71. Tai Q, Chen B, Guo F, Xu S, Hu H, Sebo B, Zhao XZ. In situ prepared transparent polyaniline electrode and its application in bifacial dye-sensitized solar cells. ACS Nano 2011; 24;5(5):3795-9.

72. Ameen S, Akhtar MS, Kim YS, Yang OB, Shin HS. Sulfamic Acid-Doped Polyaniline Nanofibers Thin Film-Based Counter Electrode: Application in Dye-Sensitized Solar Cells. Journal of Physical Chemistry C 2010;114:4760–4764.

73. Nazeeruddin, MK, Kay A, Rodicio I, Humphry-Baker R, Muller E, Liska P, lachopoulosn N, Gratzel M. Conversion of light to electricity by cis-X2bis(2,2′-bipyridyl-4,4′-dicarboxylate) ruthenium(II) charge-transfer sensitizers (X = Cl-, Br-, I-, CN-, and SCN-) on nanocrystalline titanium dioxide electrodes. Journal of the American Chemical Society 1993;115:6382–6390.

74. Li Z, Ye B, Hu X, Ma X, Zhang X, Deng Y. Facile Electropolymerized-PANI as counter electrode for low cost dye-sensitized solar cells. Electrochemistry Communications 2009;11:1768-1771.

75. Graetzel M. Photoelectrochemical cells. Nature 2011;414:338-344.

76. Kawai T, Takahashi H, Matsushima Y, Ogata T, Unuma H. Evaluation of Photocatalytic Activity of TiO2 Thin Films by Spin-Trap ESR Spectroscopy. Science of Advanced Materials, 2010; 2:74-78.

77. Thomas J, Kumar KP, Mathew S. Enhancement of Sunlight Photocatalysis of Nano TiO2 by Ag Nanoparticles Stabilized with D-Glucosamine. Science of Advanced Materials 2011;3:59-65.

78. Lee SM, Cho SN, Cheon S. Anisotropic Shape Control of Colloidal Inorganic Nanocrystals. Advanced Materials 2003;15:441-444.
79. Hyam RS, Bhosale RK, Lee W, Han SH, Hannoyer B, Ogale SB. Room Temperature Synthesis of Rutile TiO2 Hierarchical Nanoneedle Flower Morphology for Dye Sensitized Solar Cell. Journal of Nanoscience and Nanotechnology 2010;10:5894-5898.
80. Du G, Wan B, Guo Z, Shen J, Li Y, Liu H. Effect of Annealing on Electrochemical Performance of Anodized TiO2 Nanotubes for Lithium Ion Batteries. Advanced Science Letters, 2011;4:469-473.
81. Mor GK, Shankar K, Paulose M, Varghese PK, Grimes CA. Use of Highly-Ordered TiO2 Nanotube Arrays in Dye-Sensitized Solar Cells. Nano Letters, 2006;6:215-218.
82. Ameen S, Akhtar MS, Kim YS, Shin HS. Controlled synthesis and photoelectrochemical properties of highly ordered TiO2 nanorods. RSC Advances, 2012;2:4807–4813
83. Shin K, Seok SI, Im SH, Park JH. CdS or CdSe decorated TiO2 nanotube arrays from spray pyrolysis deposition: use in photoelectrochemical cells. Chemical Communication, 2010;46:2385-2387.
84. Du PA, Chen HH, Lu YC. Dye sensitized solar cells using well-aligned zinc oxide nanotip arrays. Applied Physics Letters 2006;89:253513-253615.
85. Gregg BA. Interfacial processes in the dye-sensitized solar cell. Coordination Chemistry Reviews 2004;248:1215-1224.
86. Lee S, Cho IS, Lee JH, Kim HD, Kim DW, Kim JY, Shin H, Lee JK, Jung HS, Park NG, Kim K, Ko MJ, Hong KS. Two-Step Sol−Gel Method-Based TiO2 Nanoparticles with Uniform Morphology and Size for Efficient Photo-Energy Conversion Devices. Chemistry of Materials, 2010;22:1958-1965.
87. Chakravarty R, Periasamym C. Effect of aluminium doping on structural and optoelectronic properties of sol-gel derived nanocrystalline ZnO thin film. Science of Advanced Materials 2011;3:276–283.
88. Chao CH, Chan CH, Huang JJ, Chang LS, Shih HC. Manipulated the band gap of 1D ZnO nano-rods array with controlled solution

concentration and its application for DSSCs. Current Applied Physics 2011;11:S136–S139

89. Mohanta SK, Kim DC, Kong BH, Cho HK, Liu W, Tripathy S. Optical Properties of ZnO Nanorods and Hybrid Structures Grown on p-type GaN/Sapphire and Silicon-on-Insulator Substrates. Science of Advanced Materials 2010;2:64-68. (b) Zhang H, Du N, Chen B, Li D, Yang D. Carbon Nanotube-ZnO Nanosphere Heterostructures: Low-Temperature Chemical Reaction Synthesis, Photoluminescence, and Their Application for Room Temperature NH3 Gas Sensor. Science of Advanced Materials 2009;1:13-17.
90. Akhtar MS, Khan MA, Jeon MS, Yang OB. Controlled synthesis of various ZnO nanostructured materials by capping agents-assisted hydrothermal method for dye-sensitized solar cells. Electrochimica Acta 2008;53:7869–7874.
91. Tseng YK, Hsu HC, Hsieh WF, Liu KS, Chen IC. Two-step oxygen injection process for growing ZnO nanorods. Journal of Materials Research 2003;18:2837–2844.
92. Yu WD, Li XM, Gao XD, Self-catalytic synthesis and photoluminescence of ZnO nanostructures on ZnO nanocrystal substrates, Applied Physics Letters 2004;84:2651–2658.
93. Uthirakumar P, Kang JH, Senthilarasu S, Hong CH. The different types of ZnO materials on the performance of dye-sensitized solar cells, Physica E 2011;43:1746–1751.
94. Greene LE, Yuhas BD, Law M, Zitoun D, Yang PD. Solution-Grown Zinc Oxide Nanowires. Inorganic Chemistry 2006;45:7535-7543.
95. Han H, Bach U, Cheng YB, Caruso RA. Increased nanopore filling: Effect on monolithic all-solid-state dye-sensitized solar cells. Applied Physics Letters 2007;90:213510-213512.
96. Gao YF, Nagai M, Chang TC, Shyue JJ. Solution-Derived ZnO Nanowire Array Film as Photoelectrode in Dye-Sensitized Solar Cells. Crystal Growth and Design 2007;7:2467-2471.
97. Jiang CY, Sun XW, Tan KW, Lo GQ, Kyaw AKK, Kwong DL. High-bendability flexible dye-sensitized solar cell with a nanoparticle-modified ZnO-nanowire electrode. Applied Physics Letters 2008;92:143101-143103.

98. Akhtar MS, Khan MA, Jeon MS, Yang OB. Controlled synthesis of various ZnO nanostructured materials by capping agents-assisted hydrothermal method for dye-sensitized solar cells. Electrochimica Acta 2008;53:7869-7874.
99. Umar A, Al-Hajry A, Hahn YB, Kim DH. Rapid synthesis and dye-sensitized solar cell applications of hexagonal-shaped ZnO nanorods. Electrochimica Acta 2009;54:5358-5362.
100. Ameen S, Akhtar MS, Shin HS. Growth and characterization of nanospikes decorated ZnO sheets and their solar cell application. Chemical Engineering Journal 2012;195-196:307–313.
101. Umar A, Kim SH, Kim JH, Al-Hajry A, Hahn YB. Temperature-dependant non-catalytic growth of ultraviolet-emitting ZnO nanostructures on silicon substrate by thermal evaporation process. Journal of Alloys and Compounds 2008;463:516–521.
102. Laudise RA, Kolb ED, Caporason AJ. Hydrothermal growth of large sound crystals of zinc oxide. Journal of the American Ceramic Society 1964;47:9–12.
103. Ameen S, Akhtar MS., Kim YS, Yang OB, Shin HS. Influence of seed layer treatment on low temperature grown ZnO nanotubes: performances in dye sensitized solar cells, Electrochimica Acta 2011;56:1111–1116.
104. Wagner CD, Riggs WM, Davis LE, Moulder JF, Muilenberg GE. Handbook of X-ray Photoelectron Spectroscopy, Perkin Elmer, Eden Prairie, 1979.
105. Leunga CY, Djurisic AB, Leung YH, Ding L, Yang CL, Ge WK. Influence of the carrier gas on the luminescence of ZnO tetrapod nanowires, Journal of Crystal Growth 2006;290:131–136.
106. Fan JCC, Goodenough JB. X-ray photoemission spectroscopy studies of Sn doped indium-oxide films, Journal of Applied Physics 1977;48:3524–3531.
107. Martinson ABF, Elam JW, Hupp JT, Pellin MJ. ZnO nanotube based dyesensitized solar cells Nano. Letters 2007;7:2183–2187.
108. Umar A, Singh P, Al-Ghamdi AA, Al-Heniti S. Direct growth of ZnO nanosheets on FTO substrate for dye-sensitized solar cells applications, Journal of Nanoscience and Nanotechnology 2010;10:6666–6671.

109. Al-Hajry A, Umar A, Hahn YB, Kim DH. Growth, properties and dyesensitized solar cells–applications of ZnO nanorods grown by lowtemperature solution process. Superlattice Microstructures 2009;45:529–534.

110. Ilican S, Caglar Y, Caglar M, Yakuphanoglu F. Structural, optical and electrical properties of F-doped ZnO nanorod semiconductor thin films deposited by solgel process. Applied Surface Science 2008;255:2353–2359.

111. Peiteado M, Iglesias Y, Fernandez JF, De Frutos J, Caballero AC. Microstructural development of tin-doped ZnO bulk ceramics. Materials Chemistry and Physics 2007;101:1–6.

112. Tsay CY, Cheng HC, Tung YT, Tuan WH, Lin CK. Effect of Sn-doped on microstructural and optical properties of ZnO thin films deposited by sol-gel method. Thin Solid Films 2008;517:1032–1036.

113. Yung KC, Liem H, Choy HS, 185002 enhanced redshift of the optical band gap in Sn-doped ZnO free standing films using the sol–gel method. Journal of Physics D: Applied Physics 2009;42:15002-15006.

114. Ameen S, Akhtar MS, Seo HK, Kim YS, Shin HS. Influence of Sn doping on ZnO nanostructures from nanoparticles to spindle shape and their photoelectrochemical properties for dye sensitized solar cells. Chemical Engineering Journal 2012;187: 351– 356

115. Exarhos GJ, Sharma SK. Influence of processing variables on the structure and properties of ZnO films. Thin Solid Films 1995;270:27–32.

116. Yang J, Lee J, Im K, Lim S, Influence of Sn-doping in hydrothermal methods on the optical property of the ZnO nanorods. Physica E 2009;42:51–56.

117. Sheini FJ, More MA, Jadkar SR, Patil KR, Pillai VK, Joag DS. Observation of photoconductivity in Sn-doped ZnO nanowires and their photoenhanced field emission behavior. Journal of Physical Chemistry C 2010;114:3843–3849.

118. Dolbec R, ElKhakani MA, Serventi AM, Jacques RGS. Influence of the nanostructural characteristics on the gas sensing properties of pulsed laser deposited tin oxide thin films. Sensors and Actuators

B 2003;93:566–571.

119. Ye N, Qi J, Qi Z, Zhang X, Yang Y, Liu J, Zhang Y. Improvement of the performance of dye-sensitized solar cells using Sn-doped ZnO nanoparticles. Journal of Power Sources 2010;195:5806–5809.

120. Kim MH, Suh M, Gowrishankar V, McGehee MD, Kwon YU. Confinement Effects of P3HT in Nanochannels and Their Implications for Bulk Heterojunction Solar Cells. Journal of Nanoscience and Nanotechnology 2010;10:279–284.

121. Mridha S, Basak D. ZnO/polyaniline based inorganic/organic hybrid structure: Electrical and photoconductivity properties. Applied Physics Letters 2008;92:142111–142113

122. Ameen S, Akhtar MS, Kim YS, Yang OB, Shin HS. Polyaniline/ Ga doped ZnO Heterostructure Device via Plasma Enhanced Polymerization. Microchim Acta 2011;172:471–478.

123. Umar A, Al-Hajry A, Hahn YB, Kim DH. Rapid synthesis and dye-sensitized solar cell applications of hexagonal-shaped ZnO nanorods. Electrochimica Acta 2009;54:5358–5362.

124. Monkman AP, Stevens GC, Bloor D. X-ray photoelectron spectroscopic investigations of the chain structure and doping mechanisms in polyaniline. Journal of Physics D: Applied Physics 1991;24:738-741.

125. Barr TL. Recent advances in x-ray photoelectron spectroscopy studies of oxides. Journal of Vacuum Science & Technology A 1991;93:179-185.

126. Sans JA, Segura A, Royo JFS, Barber V, Fenollosa MAH, Marí B. Correlation between optical and transport properties of Ga doped ZnO thin films prepared by pulsed laser deposition. Superlattice Microstructures 2006;39:282–290.

127. Chen Y, Kang ET, Neon KG, Lim SL, Ma ZH, Tan KL. Intrinsic redox states of polyaniline studied by high-resolution X ray photoelectron spectroscopy. Colloid and Polymer Science 2001;279:73–76.

128. Kang ET, Neoh KG, Tan KL. Polyaniline: A polymer with many interesting intrinsic redox states. Progress in Polymer Science 1998;23:277–324.

129. Mridha S, Basak D. ZnO/polyaniline based inorganic/organic hybrid structure: Electrical and photoconductivity properties. Appl. Phys. Lett. 2008;92:142111–142113.

130. Ameen S, Akhtar MS, Ansari SG, Yang OB, Shin HS. Electrophoretically deposited polyaniline/ZnO nanoparticles for p–n heterostructure diodes. Superlattice and Microstructures 2009;46:872–880.

131. Croce F, Appetechi GB, Persi L, Scrosati B. Nanocomposite polymer electrolytes for lithium batteries. Nature (London) 1998;394: 456-458.

132. Kang MS, Ahnand KS, Lee JW. Quasi-solid-state dye-sensitized solar cells employing ternary component polymer-gel electrolytes. Journal of Power Sources 2008;180:896-901.

133. Katsaros G, Stergiopoulos T, Arabatzis IM, Papadokostaki KG, Falaras P. A solvent-free composite polymer/inorganic oxide electrolyte for high efficiency solid-state dye-sensitized solar cells. Journal of Photochemistry and Photobiology A 2002;149:191-198.

134. Stergiopoulos T, Arabatzis IM, Katsaros G, Falaras P. Binary Polyethylene Oxide/Titania Solid-State Redox Electrolyte for Highly Efficient Nanocrystalline TiO2 Photoelectrochemical Cells. Nano Letters 2002;2:1259-1261.

135. Akhtar MS, Park JG, Kim UY, Yang OB. Composite electrolytes of polyethylene glycol methyl ether and TiO2 for dye-sensitized solar cells—Effect of heat treatment. Materials Chemistry and Physics 2011;127:479-483.

136. Tadayyoni MA, Gao P, Weaver MJ. Application of surface-enhanced Raman spectroscopy to mechanistic electrochemistry: Oxidation of iodide at gold electrodes. Journal of Electroanalytical Chemistry 1986;198:125-136.

137. Kubo W, Murakoshi K, Kitamura T, Yoshida S, Haruki M, Hanabusa K, Shirai H, Wada Y, Yanagida S. Quasi-Solid-State Dye-Sensitized TiO2 Solar Cells: Effective Charge Transport in Mesoporous Space Filled with Gel Electrolytes Containing Iodide and Iodine. Journal of Physical Chemistry B 2001;105:12809- 12815.

138. Akhtar MS, Chun J M, Yang OB. Advanced composite gel electrolytes prepared with titania nanotube fillers in polyethylene

glycol for the solid-state dye-sensitized solar cell. Electrochemistry Communications 2007;9:2833-2837.

139. Zhang J, Han H, Wu S, Xu S, Yang Y, Zhou C, Zhao X. Conductive carbon nanoparticles hybrid PEO/P(VDF-HFP)/SiO2 nanocomposite polymer electrolyte type dye sensitized solar cells. Solid State Ionics 2007;178:1595-1601.

140. Han H, Bach U, Cheng YB. Caruso R A. Increased nanopore filling: Effect on monolithic all-solid-state dye-sensitized solar cells. Applied Physics Letters 2007;90:213510-213412.

141. Zhang X, Yang H, Xiong HM, Li FY, Xia YY. A quasi-solid-state dye-sensitized solar cell based on the stable polymer-grafted nanoparticle composite electrolyte. Journal of Power Sources 2006;160:1451-1455.

142. Nazmutdinova G, Sensfuss S, Schrödner M, Hinsch A, Sastrawan R, Gerhard., Himmler S, Wasserscheid P. Quasi-solid state polymer electrolytes for dye-sensitized solar cells: Effect of the electrolyte components variation on the triiodide ion diffusion properties and charge-transfer resistance at platinum electrode. Solid State Ionics 2006;177:3141-3146.

143. Tu CW, Liu KY, Chien AT, Yen MH, Weng TH, Ho KC, Lin KF. Enhancement of photocurrent of polymer-gelled dye-sensitized solar cell by incorporation of exfoliated montmorillonite nanoplatelets. Journal of Polymer Science A: Polymer Chemistry 2008;46:47- 53.

144. Ito BI, de Freitas JN, De Paoli MA, Nogueira AF. Application of a Composite Polymer Electrolyte Based on Montmorillonite in Dye Sensitized Solar Cells. Journal of Brazilian Chemical Society 2008:19:688-696.

Chapter 2

Impact of K-factor and Active Current Reduction during Fault-Ride-Through of Generating Units Connected via Voltage-Sourced Converters on Power System Stability

Bernd Weise [1]

[1] Application Engineering Department, DIgSILENT GmbH, Gomaringen, Germany

ABSTRACT

This study investigates the impact of the injection of additional reactive current and of active current reduction during fault-ride-through (FRT) of generating units connected to the grid via fully rated voltage-sourced converter (VSC generating units) on the behaviour and stability of the power system. The primary focus is on voltage support, transient stability and frequency stability. The investigation is carried out via computer simulations using a fictitious transmission system, based on German grid code requirements. The K-factor of the dependence of additional reactive current with voltage deviation, the method of active current limitation during FRT and the rate of active power recovery after fault clearance are varied. Results and influences are illustrated and discussed. They show that with an increasing number of converter-connected generating units in power systems, a careful selection of the parameters becomes very important. Insufficient grid code requirements may increase the risk of system instability. Conclusions are drawn and recommendations for optimal settings with respect to future development of grid codes are provided. The mechanism of loss of synchronism of VSC generating units because of inadequate current injection (current angle instability) is explained.

INTRODUCTION

The electrical behaviour of generating units connected to the grid via a voltage-sourced converter (VSC), such as type IV wind turbines or photovoltaic systems, is determined by the rating of the converters, the design of their controllers and the selected parameter settings. The desired behaviour is defined by the applicable grid code, which influences the controller design and selection of the parameter settings. Existing grid codes (e.g. [1]) usually distinguish between power plants with synchronous generators connected synchronously to the grid (synchronous power generating modules) and power park modules, which do not have this kind of generator in synchronous operation. Power park modules usually use fully rated VSCs for power injection or doubly fed induction generators.

At the end of 2011, there was a total net generation capacity of ca. 928 GW connected to European grids [2]. About ~15% of that capacity came from solar power plants (ca. 48 GW) and wind power plants (ca. 90 GW) [2]. In some countries, the percentage of power park modules was much higher, for example, ca. 21% in Portugal, ca. 27% in Spain and ca. 34% in Germany [2]. In comparison, the highest hourly load value within the whole European Network of Transmission System Operators for Electricity (ENTSO-E) area in 2011 was ca. 530

number of VSC power plants installed. The points of common coupling (PCC) of the VSC power plants are located at the 110 kV level, with HV/MV transformers down to 20 kV (MV voltage level inside VSC power plants). The power injected by these power plants is transferred via extra 110 kV cables and 380 kV/110 kV transformers to the substations of the conventional power plants, in order to have the same resulting power flow in the transmission system with or without VSC power plants, for a better comparison during analysis of transient stability. The VSC power plants are represented by aggregated models consisting of MV/LV generating unit transformers (20 kV/0.69 kV) and low-voltage (0.69 kV) VSC generating units with detailed models of controllers and the DC circuit. Some of the VSC power plants have additional capacitor banks connected to the 20 kV level. The most important points of the dynamic behaviour of the VSC generating unit models are described in Sections 3.3 and 4.

For the results presented in this paper, ca. 50% of the power injection is provided by VSC generating units, whereas the remaining 50% is injected from classical synchronous generators.

Reactive Power Control of VSC Generating Units during Normal Operation

During normal operation, that is, if the voltage is inside its normal band, the VSC power plants are operated in Q (U) control mode, which means that the reactive power of the VSC generating units is controlled in accordance with the voltage at the PCC of the VSC power plants. This kind of voltage control is not as fast as the dynamic voltage support during FRT (as described in Section 3.3), but is much faster than switched capacitor banks, for example. The Q (U) control mode is needed to stabilise the voltage after fault clearance. Since 50% of conventional synchronous generator power plants are disconnected from the grid in the cases studied in this paper, 50% of AVRs are also missing to ensure proper voltage control. Without enabling the Q (U) mode of VSC power plants, voltage instability can occur after fault clearance, if the voltage has recovered to a point at which the VSC generating units change from failure mode (FRT behaviour as described in Section 3.3) to normal operating mode again. In such a case of voltage instability, voltage drops down again. This can result in ongoing toggling between failure mode and normal operating mode of VSC generating units, as represented in [24]. Using the Q (U) control mode avoids this problem. As an alternative, the use of additional STATCOMs or synchronous condensers would also be possible (but is not used for the results presented in this paper).

FRT Behaviour of VSC Generating Units

The behaviour of the VSC-connected generating unit models fulfil the German grid code requirements for wind turbines [13]. During faults, reactive current I_Q is injected according to (1) and (2)

$$I_Q = I_{Q0} + \Delta I_Q \tag{1}$$

$$\Delta I_Q / I_n = K.((U - U_0) \mp U_{deadband}) / U_n \tag{2}$$

In (1) and (2), I_Q is the additional reactive current during FRT, I_{Q0} is the pre-fault reactive current, I_n is the nominal current (for steady-state operation) of the generating unit, U is the voltage (positive sequence) during FRT, U_0 is the pre-fault voltage, $U_{deadband}$ is the voltage deadband as specified in [13] and U_n is the nominal voltage. Equation (2) is only valid, if the voltage U deviates from the pre-fault voltage U_0 by more than the voltage deadband $\pm U_{deadband}$. $U_{deadband}$ is usually 10% of the nominal voltage U_n [13], but it is also possible to reach an agreement with the system operator for a smaller deadband (even down to 0%) [13]. For the investigation presented in this paper, the standard voltage deadband of ±10% is used ($U_{deadband} = 0.1\ U_n$).

Voltage and current in (1) and (2) are at the LV terminals of the generating units [13]. The K-factor in this paper therefore applies to the LV terminals.

In each scenario/simulation run, all VSC generating units have the same parameter settings. The settings are varied from between simulation runs to investigate the influence on the power system:

- the K-factor for the additional reactive current during FRT in the range from 0 to 10 and
- the limitation of the active current during FRT as described below.

The active current during FRT is limited to avoid overloading of the VSC (or other equipment pertaining to the generating unit), while priority is given to reactive current

$$I_P = \min(I_{P,not\,limited}, I_{P,max}) \tag{3}$$

In (3) I_P is the limited active current, $I_{P,not\ limited}$ the non-limited active current and $I_{P,max}$ is the active current limit. The limit $I_{P,max}$ is set to one of (4)–(6). In (4)–(6), I_{max} is the short-term maximum tolerable current of the generating unit and I_Q is the reactive current during FRT

a) $$I_{P,\,max} = 0A \tag{4}$$

b) $$I_{P,max} = I_{max} - |\,I_Q\,| \tag{5}$$

c) $$I_{P,\,max} = \sqrt{I^2_{max} - I^2_Q} \tag{6}$$

In addition, an alternative FRT behaviour option which reduces both active and reactive currents to zero (I_P = 0 A, I_Q = 0 A) during FRT is also implemented in the model. This alternative behaviour is sometimes used in practice; however, it is not compliant with the German grid codes for connection to the MV, HV or EHV level.

Fig. 2 shows examples of the corresponding behaviour. In these examples, I_{max} (short-term) is 1.1 of the long-term rated apparent current of the generating unit converter, whereas the currents in the figure are shown in per unit (pu) based on the nominal active current of the generating unit. Hence, I_{max} is 1.275 pu based on nominal active current as represented in Fig. 2.

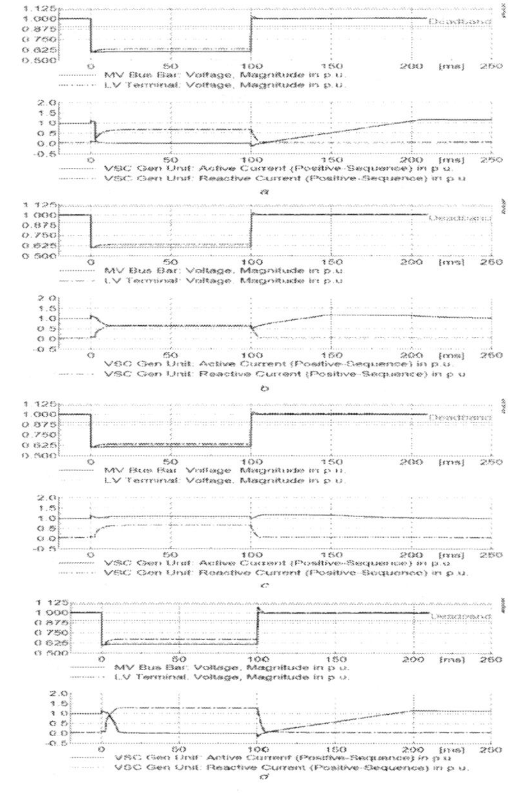

Figure 2: Example of FRT behaviour of VSC generating unit models

a K = 2, I_P according to (3) and (4)

b K = 2, I_P according to (3) and (5)

c K = 2, I_P according to (3) and (6)

d K = 5, I_P according to (3) and (6)

Apart from the variation of K-factor and active current reduction, it is possible to enable a blocking of the PLLs at low voltages. If a PLL is blocked, the frequency and voltage angle detected by the PLL are frozen, and the frequency and voltage angle output of the PLL stay constant until the blocking is released. The models are set up either without any PLL blocking at all or with blocking of the PLLs if the voltages drop below 0.40 pu Once a PLL is blocked, it is released if the voltage rises above 0.45 pu again.

Active Power Recovery of VSC Generating Units after Fault Clearance

The current controller of the VSC generating unit model has a limitation for the rate of increasing the active current. By varying the value of this limitation, it is possible to influence the rate of active power recovery after fault clearance. For the investigation presented here, the following rates are selected:

- 20%/s = 100% in 5 s.
- 100%/s = 100% in 1 s.
- 200%/s = 100% in 0.5 s.
- 1000%/s = 100% in 100 ms.

All of these settings are in-line with German grid codes [10, 13] (see Section 2). Options 3 and 4 are in-line with grid codes of Great Britain and Ireland for all voltage dips, option 2 for voltage dips with a duration of more than 140 ms [17 – 19]. Examples of these settings are shown in Fig. 4

Simulations Of Short-Circuits In the Grid

For the investigation, balanced three-phase short-circuits are simulated in the grid. The fault location for the results presented in Sections 4.1 and 4.3 of this paper (voltage support and transient stability) is on one of the tie lines at its end to Area 1, near node no. 3 as depicted in Fig. 1 (fault location SC1). Fault clearance is achieved by switching off the faulted circuit of the tie line. The fault clearing time is varied.

For the analysis of the impact on frequency stability (Section 4.2), a balanced three-phase short-circuit located on a circuit of a double line inside Area 2, in the middle of the network, is simulated (fault location SC2 in Fig. 1). the fault is cleared by switching off the faulted circuit of the double line after 100 ms.

In all simulations, the short-circuit events start at 0 s in time.

RESULTS

Results are presented for scenarios in which 50% of the load demand is covered by power plants with VSC generating units. Conventional power plants with synchronous generators supply the remaining 50% of the load demand and have reserves for primary and secondary frequency control. Surplus generators are disconnected.

Voltage Support During FRT

The resulting voltage magnitudes during short-circuit at location SC1 as described in Section 3.5 are shown in Fig. 3 for a selection of bus bars. The selected bus bars are indicated by corresponding numbers in Fig. 1. The fault location is near node no. 3. The pre-fault load flow scenario is without active power exchange and almost no reactive power exchange between the areas. The results presented are mean values of the RMS values from 30 through 80 ms after the onset of the short-circuit. For comparison, the results of the reference scenario without any VSC generating units and of a scenario in which active and reactive currents of the VSC generating units are reduced to 0 A during FRT (no voltage support) are also included in Fig. 3.

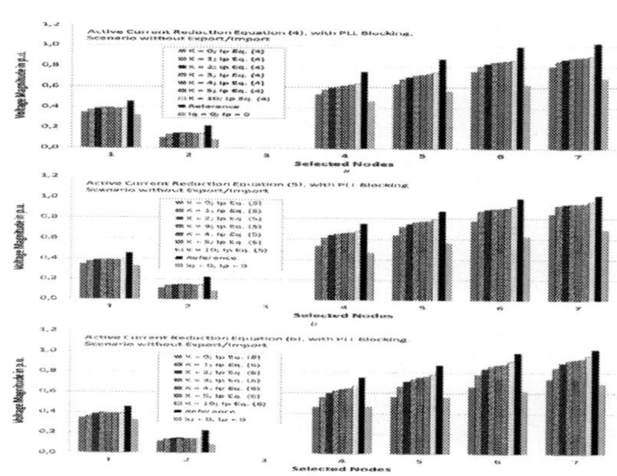

Figure 3: Voltage magnitudes at selected bus bars during short-circuit near node no. 3, additional reactive current injection with different K-factors (reference = only synchronous generators), pre-fault scenario without export/import between the areas

a I_p according to (3) and (4)

b I_p according to (3) and (5)

c I_p according to (3) and (6)

As expected, the voltage drops during short-circuit. Bus bars which are electrically nearer to the short-circuit location (for example, node nos. 1 and 2 in Area 1) have a lower voltage than bus bars further away. Voltage support of the VSC units has a positive effect (voltages are higher than without voltage support), in particular at the buses further away from the fault location. In Area 1, which is close to the fault location, with a K-factor of 2 all VSC units meet their current limits already, and hence a higher K-factor setting does not further improve the resulting voltage in that area. For the remote buses in Area 2 and especially Area 3, with increasing K-factor the voltage also increases. The biggest step of increase is from $K = 0$ to $K = 1$. In comparison, the steps for $K > 5$ are small.

In small grids voltage support is limited, because the impedances of the grids are comparably small. The correlation of impedance and possible voltage support is explained theoretically in [25]. Since the voltage becomes very low in cases of faults in small grids, all VSC generating units reach their limits with small K-factors already as in Area 1 of the fictitious transmission system (nodes no. 1 and 2 in Fig. 3). Higher K-factors have no effect.

In bigger grids VSC generating units which are far away from the fault location but still detect a voltage dip, have an influence with their parameter settings and behaviour on the remaining voltage. In the fictitious transmission grid, these are the VSC generating units in Areas 2 and 3 (node nos. 4–7 in Fig. 3).

In the cases with $K = 0$ the VSC generating units do not inject additional reactive current, but they continue to inject their pre-fault reactive current, as indicated by (1). In the simulated scenarios the VSC generating units run overexcited during steady-state operation, because the VSC power plants replace conventional power plants which run overexcited. Hence, the results for $K = 0$ show remaining voltages during the fault which are higher in most cases, even with active current reduction to 0 A according to (3) and (4) (Fig. 3 *a*, e.g. 0.75 pu at node no. 6), in comparison with reduction of active 'and' reactive currents of the VSC generating units to 0 A during FRT (Fig. 3 *a*, 0.62 pu at node no. 6).

The results show that there is also an effect of the active current injection on the voltage magnitude. If the active currents of VSC units in FRT mode are limited to 0 A, for $K > 2$ the voltage in Area 2 and Area 3 is lower (Fig. 3 *a*, e.g. 0.85 pu for $K = 5$ at node no. 6) than with active current according to (5) or (6) (Fig. 3 *b*, 0.90 pu, and Fig. 3 *c*, 0.88 pu, respectively, for $K = 5$ at node no. 6). The active current injection additionally supports voltage increase.

However, with small K-factors, especially at $K = 0$, compared with usage of (4) (Fig. 3 *a*, 0.75 pu for $K = 0$ at node no. 6), as expected voltage is higher if (5) is used (Fig. 3 *b*, 0.78 pu for $K = 0$ at node no. 6) because of the additional active current injection, but lower if (6) is used (Fig. 3 *c*, 0.66 pu for $K = 0$ at node no. 6). This contrary effect of the usage of (6) with a small K-factor is caused by loss of synchronism of VSC generating units in Areas 2 and 3. To avoid loss of synchronism, blocking of PLLs in the case of voltages below 0.40 pu has been used (see Section 5). However with (6) and $K = 0$, synchronism of VSC generating units can even get lost at higher voltages (still below normal operating range). This is because it is not possible to transfer as much active power through the grid at such voltages as the VSC generating units inject active current (locally) [25]. The reason is the reduced load demand at lower voltages and the reduced transferable active power through lines and transformers at lower voltages (voltage instability issue [4, 5]). The inadequate current injection leads to a change of frequency of the injected current of VSC generating units (see Section 5). Owing to the incorrectly synchronised injection, voltage cannot be supported effectively. With higher K-factors the active current injection is reduced, which lowers the risk of the described instability.

The best results (highest voltages) are achieved with a high K-factor and (5) or (6).

Impact on Frequency Stability

For investigation of the impact on frequency stability, short-circuits at location SC2 are simulated as described in Section 3.5 and depicted in Fig. 1. The K-factor of the VSC generating units is set to $K = 4$. The rate of increase of active power after voltage recovery is varied as described in Section 3.4. The simulations are carried out with active power reduction according to (4)–(6), as well as with reduction of active and reactive currents of VSC generating units down to 0 A during

FRT. Results are shown in Fig. 4. The active power injected by a VSC generating unit, the total active power injected by all generators into the system and the frequency are displayed.

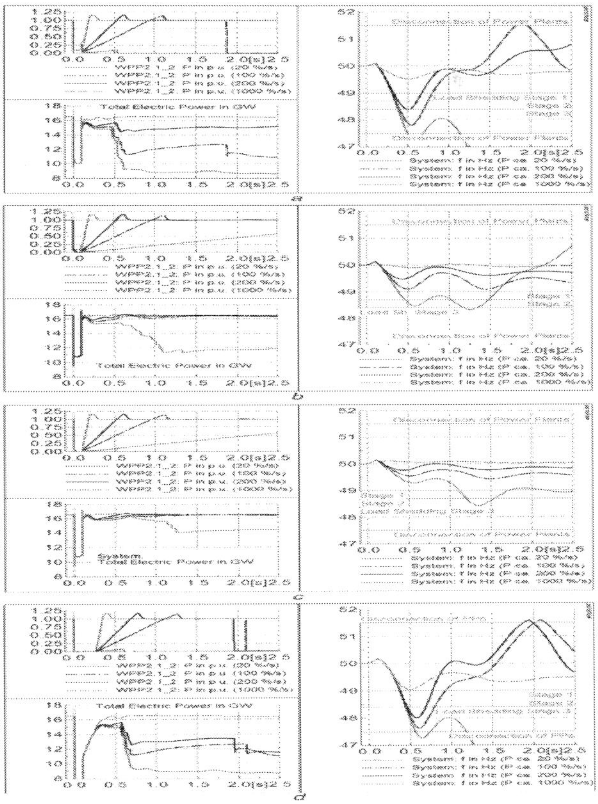

Figure 4: Frequency and active power during a fault and after fault clearance

Curves on the top left: active power injected by a VSC power plant located very near to the fault. Curves on the bottom left: total electric active power injected into the whole system. Curves on the right: electric frequency in the system

a $K = 4$, I_P according to (3) and (4)

b $K = 4$, I_P according to (3) and (5)

c $K = 4$, I_P according to (3) and (6)

d $I_Q = 0$, $I_P = 0$ during FRT

The results show, in general, that a slow rate of active power increase leads to a lack (unbalance) of generated power for 1 or 2 s which can cause the frequency to fall beyond thresholds for load shedding stages (as was already reported in [7]). The fast frequency decrease is additionally worsened by a relatively small system inertia (the inertia time constant of the fictitious transmission system with 50% synchronous generators connected is ~70% of the inertia time constant of the system with 100% synchronous generators connected).

In the case of active current limitation according to (3) with (4), an active power recovery with a rate of ca. 20%/s leads to a decrease of the frequency to below 47.5 Hz within 0.5 s (Fig. 4 *a*). Load shedding does not help to prevent such a low frequency. The frequency threshold for the disconnection of power plants is 47.5 Hz. As frequency relays are modelled with the VSC generating units, but not included in the models of the conventional synchronous generator power plants, in the simulation the power injection of the synchronous generators remains, and the curve of the total electric power does not drop to zero. Nevertheless, in reality all power plants would disconnect in this scenario and a complete black-out of the system would result.

Faster rates of active power recovery avoid disconnection of power plants, but with 100 and 200%/s the frequency still decreases into ranges of load shedding. In the curve of the total electric active power, load shedding is visible as a decrease, because the injected electric power is always in balance with consumption of loads and losses (it is the turbine power which has a lack or surplus in case of unbalances). As load shedding occurs during the increase of active power of the VSC generating units, 1 s after load shedding there is a surplus of turbine power and the frequency rises quickly to above 50 Hz. In cases of active power recovery with 100%/s, load shedding of stage 3 is executed, following which the frequency rises slightly to above 51.5 Hz within <2 s. Disconnection of some power plants is the result.

With an active current limitation according to (3) with (5) or (6) (Figs. 4 *b* and *c*), the starting point for recovery of active power injected by VSC generating units further away from the fault location is at a higher level. Consequently, the difference between power injected by generating units and power demand of loads is smaller. The frequency does not decrease dramatically; however, load shedding still occurs for the slowest simulated rate of active power recovery (20%/s).

If the voltage was not supported by VSC generating units during FRT and their active and reactive currents was reduced to 0 A, voltage recovery after disconnection of the short-circuited line is much slower. The Q (U) control mode for normal operation of the VSC power plants (Section 3.2) is frozen during FRT and the control loop of the Q (U) mode is only activated again if the voltages at the terminals of the VSC units return into the voltage deadband described in Section 3.3. As long as the voltage in the grid is below the normal operating range (compare Fig. 3), voltage recovery is only supported by the remaining synchronous generators. Hence, voltage recovery is slower and at a lot of VSC unit terminals the voltage reaches the normal operating range later. As a consequence, active power recovery of the VSC generating units starts later (Fig. 4 *d*) and from a low level (0 MW). Hence, the frequency decreases heavily in scenarios with these settings for VSC generating units. Disconnection of power plants because of under- or over-frequency after load shedding occurs with recovery rates of 20, 100 and even 200%/s.

Only with the fastest recovery rate of 100% in 100 ms (1000%/s), it is possible to avoid load shedding for all cases of active current reduction during FRT.

Please note that in the simulations presented here, no VSC generating unit disconnects during FRT. In real networks, generating units connected to the LV grid would probably disconnect (according to German grid codes they must disconnect [26]), which would cause a further lack of active power after fault clearance and hence even worst situations.

The situation in Continental Europe is probably not as severe, because the interconnected grid is bigger and has a higher system inertia than the fictitious transmission system used in this investigation. However, it should be mentioned that there is the risk that the interconnected network could split into smaller islanded areas after severe disturbances which may have a size similar to the fictitious network presented here.

To avoid frequency instability and minimise load shedding, the active current reduction during FRT should be as modest as possible and active power recovery after fault clearance should have a high rate: 200%/s or even 1000%/s is recommended based on the results presented in this paper.

Impact on Transient Stability

To analyse the impact on transient stability, the critical fault clearing time (CFCT) is evaluated for short-circuits on the tie line between Area 1 and Area 2 (location SC1 near node no. 3 as described in Section 3.5 and depicted in Fig. 1). The CFCT is the maximum duration of a short-circuit, which must not be exceeded in order to assure transient stability of the conventional synchronous generators. Results are presented in Fig. 5 for a power flow scenario with 1000 MW exported from Area 1 into Area 3 (1000 MW export scenario), and another scenario with 900 MW flowing from Area 3 through Area 2, imported by Area 1 (900 MW import scenario, see Fig. 1 for export/import directions), both without PLL blocking and with PLL blocking as described in Section 3.3. Again K-factor and limitation of active current are varied. The speed of active power recovery after fault clearance is ca. 100% in 100 ms (1000%/s). During normal operation, Q (U) control is established as discussed in Section 3.2. In contrast to the results of CFCT presented in [6], no additional STATCOMs are connected to the grid for the results presented here.

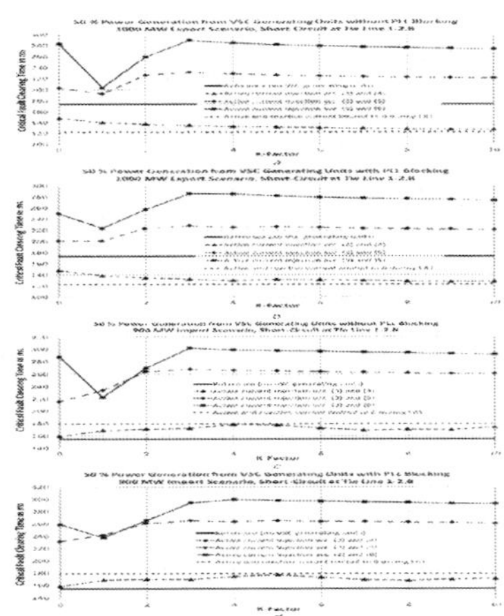

Figure 5: Critical fault clearing times for short-circuit at tie line (short-circuit location SC1). *a* 1000 MW Export Scenario, without PLL blocking, *b* 1000 MW Export Scenario, with PLL blocking below 0.4 pu voltage, *c* 900 MW Import Scenario, without PLL blocking, and *d* 900 MW Import Scenario, with PLL blocking below 0.4 pu voltage

The CFCT of the case without any VSC generating unit is taken as reference. Compared with this reference the CFCT can either increase or decrease, depending on the pre-fault load flow scenario and on the method of active current reduction during FRT. Although there is always an improvement of the CFCT in the case of the 900 MW import scenario (Figs. 5 *c* and *d*), the CFCT decreases in the 1000 MW export scenario if the active current is reduced to 0 A during FRT (Figs. 5 *a* and *b*). The latter is in-line with results of other studies with generating units either without FRT capability or without voltage support during FRT [27]. If the active current reduction during FRT is modest according to (3) with (5) or (6), the CFCT clearly improves.

The results show that for small K-factors (K = 0...3) the CFCT increases with increasing K-factor. The case $K = 0$ with (6) is an exception, in which VSC generating units that are electrically near to the fault location become unstable, as explained in detail in Section 5. This instability is avoided by blocking PLLs (see Section 5). By comparing Fig. 5 *a* with Figs. 5 *b* and *c* with Fig. 5 *d* it becomes obvious that PLL blocking influences the results for CFCT mainly for active current reduction according to (5) and (6) with small K-Factors (K = 0 and 1, a little bit at 2 and 3). $K = 0$ and $K = 1$ with (5) or (6) are the cases in which loss of synchronism of VSC generating units because of current angle instability (see Section 5) mainly occurs. With $K = 0$ and (6) even for VSC generating units with higher voltages at their terminals, at which PLLs are not blocked, loss of synchronism can occur, as was explained in Section 4.1. For K-factors equal to 4 and higher, all VSC generating units run stable. For K-factors equal to 3 and higher the CFCT is almost constant on a level which depends on the method of active current limitation. For very high K-factors the CFCT decreases slightly. For the fictitious transmission system the optimum (longest CFCT) is obtained with (6) in the range of $K = 3$ through ~5.

One of the reasons that VSC generating units can have a positive effect on transient stability is because they have no inertia and no rotating mass which can swing 'visibly' to the grid. The main reasons for the effects of the different parameter settings on the CFCT are the influence on voltage magnitude and the differences in injected active power. A higher voltage causes a higher electrical torque at the remaining synchronous generators (higher active power injection into the grid) and thus helps to lower the acceleration of synchronous machines at a closer distance to the fault location. A higher remaining voltage also keeps the load demand higher.

The reduction of active current injection of VSC units has the same effect on the synchronous generators as a (relatively) higher load demand, which also lowers the acceleration or increases the deceleration, not only during the fault but also during and after voltage recovery. However, this effect is especially important for synchronous generators at a further distance from the fault location, in areas in which the voltage is decreased only slightly and thus loads still demand power. Depending on the pre-fault load flow scenario (export or import) and the remaining voltage and load demand during the fault these generators can either accelerate or decelerate. For transient stability inside a big system, the relation of acceleration or deceleration between the generators is important. For example, in the 1000 MW export scenario, Area 2 and 3 are importing areas, in which synchronous generators decelerate slightly during the fault (reference scenario). If VSC generating units provide more active current (active current reduction acc. to (3) with (5) or (6)), they decelerate less or even accelerate and thus the maximum rotor angle deviation between Areas 1 and 2 is smaller (supporting effect for transient stability) compared with the reference scenario without VSC generating units. If the VSC generating units reduce active current considerably, synchronous generators in Area 2 decrease more, hence the max. rotor angle deviation becomes bigger, which is against transient stability, and the CFCT decreases in these cases. In the 900 MW import scenario, all synchronous generators accelerate during the short-circuit, the generators in Area 1, close to the fault, faster than the generators in Area 2 and 3 (exporting areas). If VSC generating units in Area 2 and 3 reduce their active current injection only slightly, the synchronous generators in these areas accelerate faster too (max. rotor angle deviation between Areas 1 and 2 remains smaller), if the VSC generating units reduce their active current injection greatly, synchronous generators accelerate less (max. rotor angle deviation between Areas 1 and 2 becomes larger). Hence, for transient stability of the analysed system the optimal active current limitation is according to (3) with (6).

It should be mentioned that with a much larger number of VSC generating units, issues of frequency stability become increasingly severe (because of the smaller remaining inertia in the grid) [28 , 29] and have to be solved in order to keep transient stability possible.

ANGLE INSTABILITY IN CASES OF INADEQUATE CURRENT INJECTION

In Section 4.3, it was mentioned that the case of riding through a deep voltage dip without reducing the active current requires further analysis. If a short-circuit which forces the voltage at the fault location to zero or very low values is close to the VSC generating unit, the current of the VSC flows directly into the short-circuit, the path of current flow is more or less decoupled from the rest of the grid. The ratio of the voltage at the VSC terminal and the current injected by the VSC into the (faulted) grid has to become equal with a physical quantity: the impedance between the VSC terminal and the short-circuit location. The development of the corresponding equivalent circuit is depicted in Fig. 6. In the simplified equivalent circuit, the VSC generating unit is a controlled current source injecting current into an impedance. Typical implementations of controllers and PLLs of VSC generating units are, for example, according to [15, 30]. Although a VSC generating unit is actually a controlled voltage source, it is more suitable to understand it as a current source in this context, because of the fast current controller of the VSC generating unit. The voltage at the terminal of the VSC generating unit in the simplified representation is equal to the voltage drop that the current causes at the impedance.

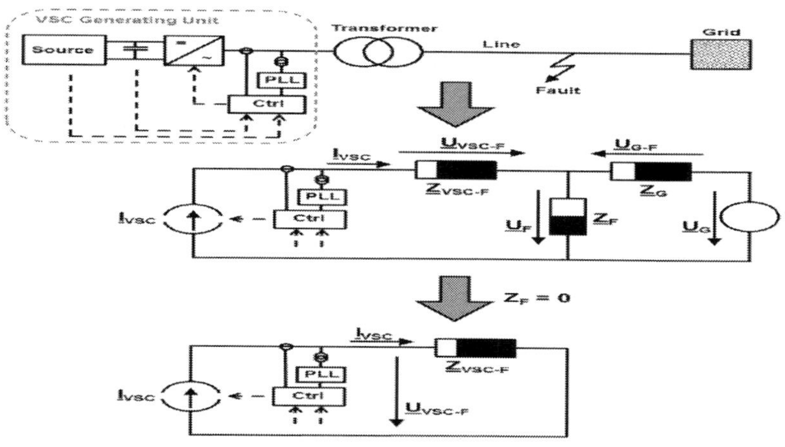

Figure 6: Equivalent circuit for explanation of current angle instability.

The angle between the voltage phasor and the current vector is equal to the impedance angle $_{Z}$. Fig. 7 shows the corresponding phasor diagrams. Please note that the coordinates of the phasor diagrams are chosen with the true current phasor as reference for the real axis, in order to locate the impedance inside the first quadrant, hence the *d*-axis of the PLL, which is the reference for the VSC controllers, is not the real axis of the diagrams. If the VSC was an uncontrolled AC voltage source, a current adequate to the impedance would flow automatically. However, as the VSC is a controlled voltage source behaving as a current source, the current controller tries to inject a current according to its current set-points (for active and reactive currents). If the resulting current angle of the current set-points fits with the impedance angle, stable operation is ensured ('current angle stability'). If the resulting current angle of the current set-points does not fit with the impedance angle, the current controller cannot find a steady-state operating point. Hence, the current angle moves and with it the local voltage angle, which is affected by the current injection.

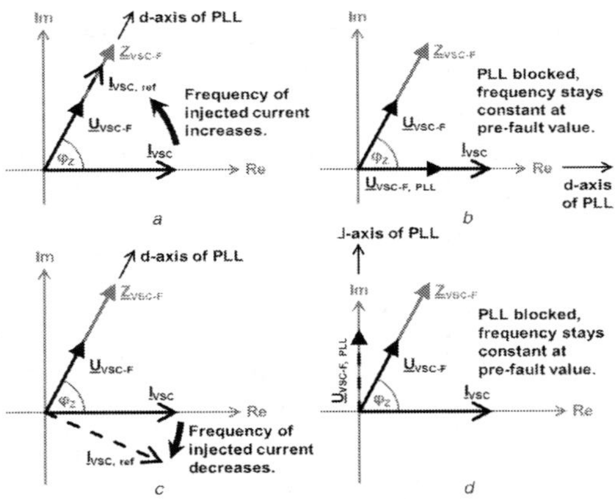

Figure 7: Phasor diagrams for explanation of current angle instability. *a* Current controller set-points to inject active current only, PLL not blocked, *b* Current controller set-points to inject active current only, PLL blocked, *c* Current controller set-points to inject reactive current only, PLL not blocked, and *d* Current controller set-points to inject reactive current only, PLL blocked.

Fig. 7 *a* represents a case in which the set-points of the current controller give a pure active current (black dashed phasor $I_{VSC,ref}$ in Fig. 7 *a*), while the impedance is assumed to be resistive–inductive.

The current controller has a deviation from its set-points as long as the current phasor does not have the same direction as the voltage phasor. Hence, the controller changes the current's phase angle to bring it onto the same phase angle as the voltage. However, because of the impedance and laws of physics, the angle between the two phasors is always the impedance angle. Hence, whenever the angle of the current phasor changes, the angle of the voltage phasor changes as well. The PLL of the VSC generating unit detects the changing voltage angle. The output of the PLL is an input of the current controller. Hence, the current controller still has the same deviation and continues to change the current's angle. As a result the local system of voltage and current rotates with a frequency higher than the frequency of the external surrounding grid. The VSC generating unit has lost synchronism with the grid frequency. The resulting frequency of the current injected by the VSC depends on the design and parameter settings (gain and limits) of the PLL and the current controller.

Figs. 8 *a* and *b* show simulated curves (over time) of the same example. The VSC injects current into a short-circuit with a fault impedance $Z_F > 0\ \Omega$, which is located near the MV terminal of the generating unit transformer (Fig. 6 top and middle). The transformer and the fault location are connected via a resistive–inductive grid impedance (the line in Fig. 6 top), the transformer impedance is mainly inductive. The set-points of the VSC current controller give a pure active current at the LV terminal in the simulated example ($K = 0$, $I_Q = 0$ A). The results in Fig. 8 *a* show that the resulting injected current at the LV terminal is mainly reactive (Fig. 8 *a* bottom left), and that the voltage angle (150° phase shift between MV and LV sides of the unit transformer) moves during the voltage dip (Fig. 8 *a* top right). The frequency of the injected current during FRT is 60 Hz (instead of 50 Hz grid frequency) because in the example 60 Hz is set up as the upper limit of the PLL (Fig. 8 *b*, bottom left). The separation into active and reactive current components at the LV terminal of the VSC generating unit (Fig. 8 *a* bottom left) shows an additional small 10 Hz beat, because the voltage at the LV terminal oscillates with a frequency of 60 Hz which has an additional small 10 Hz beat. The separation into active and reactive currents is always in relation to the local voltage angle at the given point in the grid. At the LV terminal of the VSC generating unit, the current and the voltage have almost the same frequency (60 Hz) because the voltage is dominated by the voltage drop which the 60 Hz current causes when flowing through the unit transformer. The current is mainly reactive because of the transformer's reactance, which is the biggest impedance in the path. The separation into active and reactive components of the current shows smooth curves. At the MV terminal of the transformer, however, the voltage neither oscillates with 60 Hz nor with 50 Hz. The voltage at this point is the result of the voltage drop (which the 60 Hz current of the VSC generating unit causes when flowing through the impedance between the MV terminal and the fault location), plus the remaining voltage at the fault location (mainly 50 Hz because of the short-circuit current of the grid). The separation of the current into its active and reactive components at the MV terminal (i.e. in relation to the voltage at the transformer MV terminal, Fig. 8 *a* bottom right) shows a well-developed 10 Hz beat, which is caused by the difference between the current's frequency (60 Hz) and the grid's frequency (50 Hz). Other generators or loads which may be connected to the same part of the grid between the point of coupling of the VSC generating unit and the fault location may be affected by this incorrectly synchronised injected current and its corresponding voltage drop.

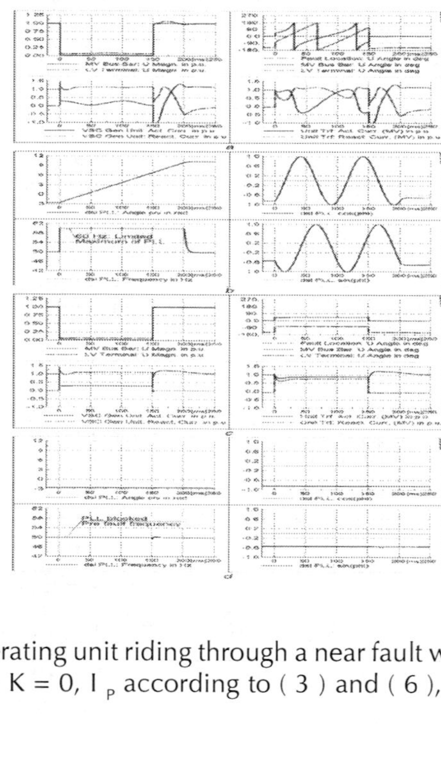

Figure 8: VSC generating unit riding through a near fault with fault impedance of nearly 0 Ω with K = 0, I_P according to (3) and (6), example of current

angle instability/stability. *a* PLL not blocked: voltage magnitudes and angles, active and reactive currents during current angle instability (loss of synchronism), *b* PLL not blocked: PLL behaviour during current angle instability (loss of synchronism), *c* PLL blocked below 0.4 pu voltage: voltage magnitudes and angles, active and reactive currents with current angle stability, and *d* PLL blocked below 0.4 pu voltage: PLL behaviour with current angle stability.

Fig. 7 *c* represents a case in which the set-points of the current controller give a pure reactive current (black dashed phasor $I_{VSC,ref}$ in Fig. 7 *c*). Again the current controller has a deviation from its set-points, this time in the opposite direction of the current angle, because the impedance is resistive–inductive, not purely inductive. Owing to the mismatch of the angles, in this case the local system of voltage and current rotates (oscillates) with a frequency lower than the frequency of the external surrounding grid.

This effect of instability might be called 'current angle instability', as an inadequate current angle of the controller set-points causes the loss of synchronism of the VSC unit. It is not rotor angle instability, because it is not related to the inertia of a swinging rotating mass, and therefore should not be called transient instability as defined according to [4 , 5].

This kind of instability was already reported in a different context for connection points with very low short-circuit ratios [31] and partly analysed in [25]. A general explanation of loss of synchronism caused by current angle instability and derivation of the according stability limits in cases of a remaining voltage at the fault location is given in [32, 33].

The resulting effect of the current angle instability is a loss of synchronism and a loss of controllability of the current angle and therefore of active and reactive current components. If the PLL is designed and set up properly, as soon as the voltage recovers, the PLL and current injection can quickly synchronise with the grid voltage again (see Figs. 8 *a* and *b*), as the VSC has no inertia.

A simple way to avoid current angle instability is to block the PLL at low voltages. If the voltage drops below a given threshold, the PLL freezes its detected frequency and voltage angle. With a blocked PLL, it is not possible to detect the true voltage angle, hence it also is not possible to control the current angle, but the current snaps to the appropriate current angle fitting with the impedance automatically (determined by the laws of physics), as depicted in phasor diagrams in Figs. 7 *b* and *d*. As a possible change of the voltage angle is not detected (the PLL is blocked and therefore 'blind'), the current controller will 'think' it injects the current with the angle according to its set-points (although it does not). The deviations between current set-points and measured values become zero (as feedback of the true voltage angle is blocked, the current controller obtains a voltage angle from the PLL that fits in the end, although the system of true voltage phasor and true current phasor may have a different direction), and the current injected by the VSC remains synchronised with grid frequency (as long as there is a small remaining voltage which links to the rest of the grid, otherwise the injected current simply remains with the pre-fault frequency). Simulated curves of voltages, currents and behaviour of a PLL which is blocked at low voltages are depicted in Figs. 8 *c* and *d*. The curves show that the VSC generating unit remains synchronised with the grid frequency.

RECOMMENDATIONS FOR FUTURE GRID CODES

As active current and reactive current injection during faults have a clear impact on system behaviour, especially on the remaining voltage, on frequency and transient stability and local current angle stability, precise requirements should be stated in every grid code for connection of generators.

Additional reactive current should be injected to support the voltage. Active current should be limited only modestly to support transient stability and to ensure frequency stability. Both additional reactive current injection and active current limitation have to be done depending on the prevailing voltage. Appropriate rules need to be written in grid codes precisely (for example, as suggested here with (5) or (6) or as proposed in [14]). Active power should be increased after fault clearance (voltage recovery) with a high recovery rate of 200%/s up to 1000%/s to ensure frequency stability and to avoid load shedding after short-circuits.

There are optimal ranges of settings which might depend on the size, topology, voltage levels and other characteristics of the power system.

Hence, individual studies are necessary to find optimal settings. For the fictitious transmission grid used in the investigation presented in this paper, a K-factor of 4, an active current limitation according to (3) with (5) or (6) and an active power recovery rate of 1000%/s give very good results in all analysed scenarios and satisfy all stability issues discussed in this paper. The higher the K-factor, the better the voltage profile during short-circuits.

SUMMARY

On the basis of existing requirements for FRT behaviour of generating units, dynamic simulations of short-circuits with different parameter settings of the VSC generating units were carried out using a model of a fictitious transmission grid. Parameters for additional reactive current injection, active current reduction and rate of active power recovery were varied and the influences on the voltage profile during the fault and on power system stability were analysed. The results show that active and reactive current injections have an impact, and that there are optimal ranges of parameter settings. Besides the classical stability criterion for electric power systems, an instability of the current and voltage angles in operation of VSC generating units can occur because of loss of synchronism caused by an inadequate current injection (current angle instability). To reduce the risk of current angle instability, it is recommended to block PLLs at low voltages. During FRT injection of additional reactive current dependent on the voltage deviation with a high K-factor (e.g. 3 through 5 would be optimal for transient stability of the fictitious transmission system investigated in this paper) and a modest reduction of the active current dependent on the reactive current injection or on the voltage deviation are recommended. The rate of active power recovery after fault clearance must be high (200–1000%/s recommended) to ensure frequency stability and to avoid load shedding following short-circuits. Precise corresponding requirements and ranges of settings which are optimal for the individual power system should be included in grid codes.

REFERENCES

1. ENTSO-E: ' *Network code for requirements for grid connection applicable to all generators'* (NC RfG) (European Network of Transmission System Operators for Electricity ENTSO-E, Brussels, Belgium, 8 March 2013) .
2. ENTSO-E: ' *Statistical yearbook 2011'* (European Network of Transmission System Operators for Electricity ENTSO-E, Brussels, Belgium) .
3. CIGRÉ Joint Working Group C1/C2/C6.18: ' *Coping with limits for very high penetrations of renewable energy'* (CIGRÉ Technical Brochure, No. 527, February 2013, ISBN 978-2-85873-220-3) .
4. KundurP.: ' *Power system stability and control'* (McGraw-Hill, Inc., New York etc., 1994), ISBN 0-07-035958-X .
5. KundurP.PaserbaJ.AijarapuV. et al. IEEE/CIGRE Joint Task Force on Stability Terms and Definitions: ' Definition and classification of power system stability', *IEEE Trans. Power Syst.*, 2004, 19, (2), pp. 1387–1401 .
6. WeiseB.: ' *Optimal K-factor and active current reduction during fault-ride-through of converter-connected generating units for power system stability* '. 12th Wind Integration Workshop, London, UK, October 2013 .
7. WeiseB.: *'Beeinflussung der Netzstabilität durch die Wirkstromreduktion im LVRT-Modus von Stromrichter-Erzeugungseinheiten und den Gradienten der Wirkleistungssteigerung nach Fehlerklärung' ('impact of active current reduction in LVRT mode of generating units connected via converters and of active power ramping speed after fault clearance on network stability')* (ETG-Kongress, Berlin, Germany, November 2013) .
8. dena-Netzstudie: ' *Energiewirtschaftliche Planung für die Netzintegration von Windenergie in Deutschland an Land und Offshore bis zum Jahr 2020'* (Deutsche Energie-Agentur GmbH (dena), Köln, Endbericht, 2005) .
9. UCTE: ' *Operation handbook'*, Appendix 1: 'Load frequency control and performance' (Union for the Co-ordination of Transmission of Electricity UCTE, 2004) .

10. VDN: ' *TransmissionCode 2007'* (Verband der Netzbetreiber VDN e.V. beim VDEW, Berlin, Germany, Version 1.1 , August 2007) .

11. BDEW: ' *Technische Richtlinie Erzeugungsanlagen am Mittelspannungsnetz – Richtlinie für Anschluss und Parallelbetrieb von Erzeugungsanlagen am Mittelspannungsnetz* ' (BDEW Bundesverband der Energie- und Wasserwirtschaft e.V., Berlin, Germany, June 2008) .

12. BDEW: ' *Regelungen und Übergangsfristen für bestimmte Anforderungen in Ergänzung zur Technischen Richtlinie: Erzeugungsanlagen am Mittelspannungsnetz – Richtlinie für Anschluss und Parallelbetrieb von Erzeugungsanlagen am Mittelspannungsnetz* ' (BDEW Bundesverband der Energie- und Wasserwirtschaft e.V., Berlin, Germany, 1 January 2013) .

13. 'Verordnung zu Systemdienstleistungen durch Windenergieanlagen (Systemdienstleistungsverordnung – SDLWindV)' ('ordinance on system services by wind energy plants') (Bundesgesetzblatt Jahrgang 2009 Teil I Nr. 39, Bonn, Germany, 2009).

14. P.O. 12.2: ' Instalaciones conectadas a la red de transporte y equipo generador: requisitos minimos de diseño, equipamiento, functionamiento, puesta en servicio y seguridad', Propuesta (Draft), Septiembre de 2010 .

15. TeodorescuR.LiserreM.RodríguezP.: ' *Grid converters for photovoltaic and wind power systems*' (John Wiley & Sons, 2011), ISBN 978-0-470-05751-3 .

16. FGW: ' *Technische Richtlinien – Teil 4: Anforderungen an Modellierung und Validierung von Simulationsmodellen der elektrischen Eigenschaften von Erzeugungseinheiten und –anlagen* ' ('TR4') (FGW e.V. – Fördergesellschaft Windenergie und andere Erneuerbare Energien, Berlin, Revision 6, 1 May 2013) .

17. ' *The grid code*', Issue 5, Revision 6 (National Grid Electricity Transmission plc, UK, 13 December 2013) .

18. ' *Guidance notes – power park modules*' (National Grid Electricity Transmission, Issue 3 , UK, September 2012) .

19. ' *EirGrid grid code*', Version 5.0 (Republic of Ireland, 7 October 2013) .

20. OedingD.OswaldB.R.: ' *Elektrische Kraftwerke und Netze'* (Springer Verlag, Berlin, Heidelberg, 2004, 6th edn.), ISBN 3-540-00863-2 .

21. ConcordiaC.IharaS.: ' Load representation in power system stability studies', *IEEE Trans. Power Appar. Syst.*, 1982, PAS-101, (4), pp. 969–977 (doi: 10.1109/TPAS.1982.317163) .

22. HandschinE.ReißlingT.: ' Frequenz- und Spannungsabhängigkeit von Verbrauchergruppen und Verbraucherteilnetzen'. VDI Berichte 582: Wirkleistungs- und Blindleistungssekundenreserve, VDI Verlag, Düsseldorf , 1986, pp. 25–52.

23. NellesD.: ' Bedeutung der Spannung- und Frequenzabhängigkeit von Lasten in Netzplanung und Netzbetrieb', etz-Archiv Bd. 7 (1985) H. 1 , pp. 11–15.

24. BoemerJ.C.GibescuM.van der MeijdenM.RawnB.G.KlingW.L.: ' Network fault response of wind power plants in distribution systems during reverse power flows (Part II)'. 12th Wind Integration Workshop, London, UK, October 2013.

25. ErlichI.ShewaregaF.EngelhardtS.KretschmannJ.FortmannJ. KochF.: ' Effect of wind turbine output current during faults on grid voltage and the transient stability of wind parks'. IEEE PES General Meeting, Calgary, Canada, July 2009.

26. VDE-AR-N 4105: ' *Technische Mindestanforderungen für Anschluss und Parallelbetrieb von Erzeugungsanlagen am Niederspannungsnetz'* (VDE Verband der Elektrotechnik Elektronik Informationstechnik e.V., Frankfurt am Main, Berlin, 2011) .

27. BoemerJ.C.van der MeerA.A.RawnB.G. et al.: ' Network fault response of transmission systems with active distribution systems during reverse power flows'. Tenth Int. Workshop on Large-Scale Integration of Wind Power into Power Systems as well as on Transmission Networks for Offshore Wind Power Plants, Aarhus, Denmark, October 2011.

28. ' All island TSO facilitation of renewables study', EirGrid, SONI. Available athttp://www.eirgrid.com/renewables/facilitationofrenewables/, accessed January 2012 .

29. O'SullivanJ.KennedyA.BoemerJ. et al.: ' Maximising the real time system penetration of windfarms on the Ireland and Northern

Ireland power system'. Ninth Int. Workshop on Large-Scale Integration of Wind Power into Power Systems as well as on Transmission Networks for Offshore Wind Power Plants, Québec City, Canada, October 2010.

30. DIgSILENT PowerFactory, Technical Reference Documentation: ' *Phase measurement device*' (DIgSILENT GmbH, Gomaringen, Germany, 2013) .

31. DiedrichsV.BeekmannA.AdloffS.: ' Loss of (angle) stability of wind power plants – the underestimated phenomenon in case of very low short circuit ratio'. Tenth Int. Workshop on Large-Scale Integration of Wind Power into Power Systems as well as on Transmission Networks for Offshore Wind Power Plants, Aarhus, Denmark, October 2011.

32. GöksuÖ.: ' Control of wind turbines during symmetrical and asymmetrical grid faults'. PhD thesis , Aalborg University, Department of Energy Technology, Aalborg, Denmark, December 2012.

33. GöksuÖ.TeodorescuR.BakC.L.IovF.KjærP.C.: ' Instability of wind turbine converters during current injection to low voltage grid faults and PLL frequency based stability solution', *IEEE Trans. Power Syst.*, 2014, 29, (4), pp. 1683–1691 (doi: 10.1109/TPWRS.2013.2295261) .

Chapter 3

Enhancing the Light Harvesting Capacity ofthe Photoanode Films in Dye-Sensitized Solar Cells

Xiang-Dong Gao[1], Xiao-Min Li[1], and Xiao-Yan Gan[1]

[1] State Key Lab of High Performance Ceramics and Superfine Microstructures, Shanghai Institute of Ceramics, Chinese Academy of Sciences, Shanghai, P. R. China

INTRODUCTION

Dye-sensitized solar cell (DSC) has been receiving continuous academic and industrial attention as a potential low-cost, clean, and renewable energy source, since its inception in 1985 (Desilvestro et al., 1988; Regan et al., 1991; Gao et al., 2008; Hagfeldt et al., 2010; Yella et al., 2011). The DSC is the only photovoltaic device that uses molecules to absorb photons and converts them to electric charges without the need of intermolecular transport of electronic excitation. It is also the only photovoltaic device that separates two functions of light harvesting and charge-carrier transport, mimicking the photosynthesis found in green leaves. The primary photon-to-electron conversion process in DSC occurs at the oxide/dye/electrolyte interface, functioning at a molecular and nano scale. There exist many complicated optical, electrical, and chemical processes during the light-to-electric conversion process in DSC, including the light absorption of dye molecules (producing electrons and leaving the dyes in their oxidized states), the electron injection from dyes to the conduction band of semiconductor (e.g., TiO_2), the percolation of electrons through the mesoporous semiconductor network toward the bottom electrode, the recombination of photo-excited electrons in the porous electrode with either oxidized dyes or acceptors in the electrolyte, and the regeneration of dye molecules by iodides in the electrolyte etc. For an efficient DSC device, while above processes should be kept in a complicated and delicate balance, the high light-harvesting, the rapid electron transport, and the minimum electron-hole recombination's are essential. In special, the light-harvesting efficiency of the photoanode is the most important and indispensable factor for the high-efficiency DSC, which is mainly related to the molar extinction coefficient of the sensitizer, the dye-loading capacity of the porous electrode, and the optical path of the incident light in the electrode.

At present, the record efficiency of DSC has reached 12.03% (Yella et al., 2011) after the first breakthrough of M. Grätzel in 1991 and through the continuous efforts of the scientists for more than 20 years. With an aim to further improve the conversion efficiency of DSC to 15% or higher, recently, intensive efforts have been directed to enhance the functions of the major components in DSC, focusing on mainly three aspects, i.e., the ordered nanoporous electrode, the novel organic or quantum-dot sensitizer, and the new redox electrolyte system (Hagfeldt et al., 2010; Tetreault et al., 2012). During this journey toward high-efficiency DSC, the light-harvesting capacity of the photoanode is expected to play a critical and indispensable role by spanning the mesoporous electrodes utilizing varied nano/meso materials and the sensitizers with high molar extinction coefficient or quantum dots, and by connecting the optical functions of the photoanode with its electrical response together.

There are mainly two components concerning the light-harvesting property of DSC, the dye and the mesoporous electrode which supports the sensitizer. While the development of the novel sensitizers with higher molar extinction coefficient and better response to near-infrared wavelength has been long studied since the first breakthrough of DSC in 1991 (Nazeeruddin et al., 2001; Grtäzel et al., 2009;Yum et al., 2011), the significance of the light scattering and/or reflection of the mesoporous electrode on the light-harvesting properties is only recognized in recent years. Accompanying the rapid advance of the ordered or hybrid photoanode materials, novel semiconductor nanostructures or new concepts are widely used to construct new type photoanodes, including the plenary optical waveguide nanowire enhancing the multiple internal reflections (Wei et al., 2010), the plasmonic core-shell metal-insulator nanoparticles enhancing the near-infrared absorption (Brown et al., 2011), the dual-function mesoporous TiO_2 structures increasing both the dye-loading and the optical scattering effects (Koo et al, 2008; Huang et al., 2010), and so on. In view of the wide coverage of the optical and photoelectrical nanostructures involved in these studies and their rapid progresses, it is great valuable to make a comprehensive and in-depth review on the current status of the light-harvesting capacities of the photoanode in DSC.

The chapter will start with a brief description of the basic concept of the light-harvesting efficiency (LHE), and then give a review on five typical branches representing the significant advances in this area, including

1. the mesoporous photoanodes with high surface area,
2. the hierarchically nanostructured photoanodes,
3. the dual-function scattering layer on the top of nanocrystalline (nc) electrode,
4. the plasmonic photoanodes, and
5. the photonic crystal photoanode and others.

The basic principles of these novel nanostructures/ methods enhancing the light-harvesting capacity of DSC, together with their mutual effects on the electrical and photoelectrochemical properties of the nanoporous electrode, will be discussed in detail. Based on the in-depth analysis of literature and the authors' experience, a perspective will be presented, shedding a light on the research road in near future.

THE LIGHT HARVESTING EFFICIENCY (LHE)

The typical DSC device is a sandwich type electrochemical cell, as seen from Figure 1, with the dye molecules absorbing the incident sunlight, and the framework of TiO_2 mesoporous electrode transporting the photo-excited electrons generated from dye molecules to the bottom electrode. Therefore, the semiconductor porous electrode, or the photoanode film, is the kernel component in DSC, undertaking two major functions: supporting dye molecules, and transporting photo-excited electrons. To harvest the sunlight as far as possible, the photoanode film has to possess a high internal surface area, to guarantee the massive uptake of dye molecules. The cell performance of DSC, that is, the solar-to-electric conversion efficiency, is determined by the short-circuit photocurrent (J_{sc}), open-circuit photovoltage (V_{oc}), and fill factor (*FF*) under a definite intensity of light such as the AM 1.5 solar spectrum. J_{sc}, the critical parameter affecting the power conversion efficiency (PCE) of DSC most, can be increased by raising the light-harvesting efficiency (LHE), which is defined by Equation 1,

$$\text{LHE}(\lambda) = 1-10^{-\tilde{A}\sigma(\lambda)} \tag{1}$$

where Γ is the number of moles of sensitizer per square centimeter of projected surface area of the film, and σ is the absorption cross section in units of cm^2/mol obtained from the decadic extinction coefficient (units of $M^{-1}\ cm^{-1}$) by multiplication with 1000 cm^3/L (Nazeeruddin et al., 1993). The LHE is directly determined by the surface concentration of the dyes in the film, and the molar extinction coefficient of dye. The LHE, together with the quantum yield of charge injection (φ_{inj}) and the efficiency of collecting the injected charge at the back contact (η_c), determines the incident monochromatic photon-to-current conversion efficiency (IPCE), defined as the number of electrons generated by light in the external circuit divided by the number of incident photons (Equation 2).

$$\text{IPCE}(\lambda) = \text{LHE}(\lambda)\varphi_{inj}\eta_c \tag{2}$$

Previous studies proved that φ_{inj} is a wavelength independent parameter, and almost all of photons absorbed by the sensitizer are quantitatively converted to the conduction band electrons (Nazeeruddin et al., 1993). Meanwhile, the time-resolved laser photolysis experiments showed that the injected electrons can percolate without significant loss through the network of interconnected particles present in the nc-TiO_2 film, and the back reaction is relatively slow, indicating that the majority of the injected charges is able to reach the back contact, and the influence of η_c on IPCE is minor (Regan et al., 1990). So the wavelength-depedent LHE has predominant effects on the IPCE and the PCE of DSC.

In essence, the LHE of DSC is the electrical response of the photovoltaic device to the solar spectrum projected on earth. It is predominantly related to there factors, the wavelength dependent light-absorbing capacity of the dye molecules, the quantity of the dye molecules adsorbed on the porous electrode, and the optical path of the incident light traveled in the electrode (determining the collision numbers of the light with the dye molecules). As a wavelength-dependent parameter, the LHE is usually evaluated by the IPCE spectrum. Figure 2 gives the IPCE spectrum of a typical DSC with nc-TiO_2 electrode, N719 dye and I_2/I_3^- electrolyte system, and its comparison with the standard solar irradiation spectrum (ASTM G173). To characterize the response of DSC to the solar irradiation, the IPCE spectrum can be divided into four wavelength zones, i.e., the strong absorption zone between 500-550 nm, two moderate absorption zones between 400-500 nm and 550-700 nm, and the weak absorption zone above 700 nm. While the broadband or near-infrared absorption sensitizers are devised to harvest the sun light beyond 700 nm, the modulation of the optical path of the incident light by introducing scattering or reflection is mostly used to improve the light harvesting in two moderate absorption zones. From the viewpoint of the photoanode film (i.e., the porous electrode), there are several measures to improve the LHE, including

1. increasing the internal surface area of the electrode, which has the potential to improve the dye-loading capacity of the photoanode over a specific film thickness and area;
2. increasing the optical path of the incident light in the electrode, by introducing scattering centers in the bulk film, by introducing the scattering layer on the top of nanocrystalline electrode, and

by constructing hierarchical structures possessing strong light scattering effects; and

3. enhancing the absorption of dye molecules by introducing plasmonic metal-semiconductor structures.

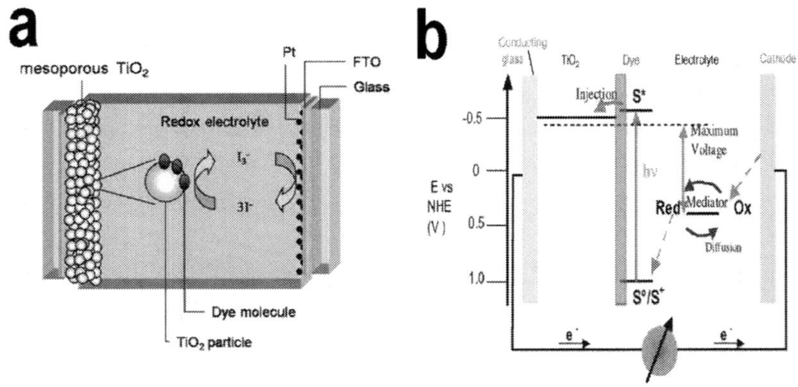

Figure 1: a) Schematic illustration of dye-sensitized solar cells (Hagfeldt, et al., 2010). b) Working principle of a typical DSC employing an iodide/triiodide-based redox electrolyte and N719 as a sensitizer (Grätzel, 2009).

In current days, the nanocrystallite electrode is confronting the maximum photocurrent theoretically achievable with the Ru-based sensitizers (e.g. N719 dye), and the quest for the novel nanostructured photoanode has becoming prosperous (Zhang et al., 2011; Tetreault et al., 2012). Among numerous innovative nanostructured photoanodes, a considerable amount of studies are directed to improve the LHE of the photoanode, in view of the moderate surface area, the moderate electron transporting property of nc-TiO_2 electrode, and the intrinsically low-surface-area nature of the traditional submicron scattering layer. In the following sections, we will introduce five typical research branches in this area, to provide readers with a comprehensive and in-depth overview on the development of the LHE studies in DSC.

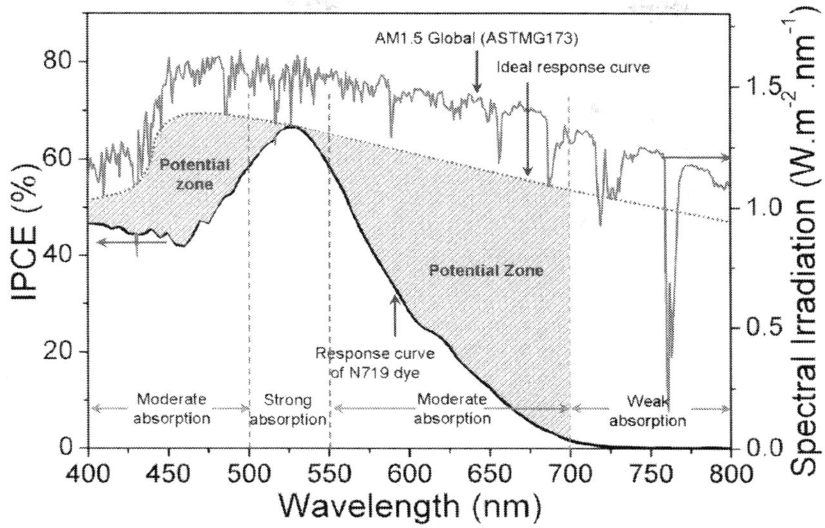

Figure 2: The light-harvesting capacity of N719 sensitized nc-TiO_2 photoanode in DSC and its comparison with AM 1.5 Global solar spectrum.

MESOPOROUS PHOTOANODES WITH HIGH SURFACE AREA

It is essential for the photoanode in DSC to possess a high internal surface area, to adsorb sensitizer as much as possible. Typically, the hydrothermal nc-TiO_2 with the particle size of 15-30 nm is used to fabricate the porous electrode, which usually exhibits the BET surface area of 50-100 m^2/g. While these nanocrystallite photoanodes perform fairly well and occupy the highland of DSC with the best efficiency, their moderate dye-loading capacity motivate researchers to seek other materials with higher surface area. In fact, there are several materials which do possess much higher surface area than nc-TiO_2 electrode, such as the mesoporous materials, aerogel, and metal-organic framework (MOF), whose surface area usually range from several hundred to three thousand m^2/g. It is greatly intriguing to construct photoanodes using these high-surface-area materials, which is expected to remarkably improve the dye-loading capacity of the photoanode and the PCE of DSC in case a good charge transport can be maintained in these mesoporous structures. Up to date, mesoporous TiO_2 in the form of powder, sphere, bead, or film, and aerogel have been used to synthesize the photoanode, which usually exhibited the surface area of 100-1100 m^2/g. Here we present some typical mesoporous photoanodes prepared by the high-surface-area TiO_2 mesostructures and TiO_2 aerogel.

Mesoporous powders possessing a large volume of mesopores and irregular micron/ submicron secondary particles are the mostly-used materials in constructing the mesoporous photoanodes. The template methods are the common strategy to prepare these mesoporous powders. For example, T. K. Yun et al. prepared two different mesoporous TiO_2 powders, using a soft template of tri-block copolymer and a hard template of mesoporous $ZnO/Zn(OH)_2$-composite, respectively (Yun et al., 2011). Both mesoporous TiO_2 possessed the same high surface area (BET ~210 m^2/g) but with different pore sizes of 6.8 (for soft template) and 3.0 nm (for hard template). Different photovoltaic performances were observed. While the photoanode using the mesoporous TiO_2 having larger pores showed the PCE of 6.71%, higher than P25 TiO_2 nanopowders based photoanode (η= 5.62%), only half the performance (3.05%) was observed for the mesoporous TiO_2 having small pores. The work indicated that, a suitable selection of the pore sizes was important for mesoporous materials used in DSC, and a too small pore size would inhibit the diffusion of dye molecules through the pores, thus greatly reducing the uptake of the dye molecules.

Beside the template method, the external fields such as the microwave irradiation can also be very useful in synthesizing mesoporous TiO_2 powders. C. H. Huang et al. reported the rapid synthesis of mesoporous TiO_2 via a microwave-assisted hydrothermal route using water-soluble titanium citrate complexes as the precursor, and obtained TiO_2 powders showed the surface area of 217–323 m^2/g, and the pore diameter of 5.8–6.9 nm (Huang et al., 2011). The photoanode based on this mesoporous TiO_2 exhibited the highest conversion efficiency of 7.1% (active area: 0.25 cm^2). Also, the cell exhibited good long-term stability, and the PCE of a 1 cm^2 device decreased merely from 4.8% to 4.3% after being stored for more than 490 h.

For most photoanodes using the high-surface-area mesoporous materials, it is required to prepare the mesoporous powders first, and the final electrode is obtained via the doctor-blade method using the mesoporous paste prepared from the mesoporous powders. This is a relatively complicated process. In contrast, W. Chen et al. developed a facile one-step preparation of crack-free thick ordered mesoporous TiO_2 films, via the combination of "doctor blade" technique and the evaporation induced self-assembly method (Chen et al., 2007). By employing the as-synthesized mesoporous film (7 μm in thickness) as the photoanode, a PCE of 6.53% was obtained at 30 mW/cm^2 light intensity, illustrating the potential of this simple strategy in constructing powerful DSC devices.

While the mesoporous materials of pure TiO_2 give a promising potential to improve the LHE, the introduction of other oxides (e.g., Al_2O_3, SiO_2) in the mesoporous structure of TiO_2 opens an alternative route to enhance the LHE of DSC. As an example, J. Y. Kim et al. prepared highly ordered mesoporous Al_2O_3/TiO_2 via a sol-gel reaction and evaporation-induced self-assembly route using Pluronic P123 as the structure-modifying reagent (Kim et al., 2011). Obtained mesoporous powders possessed an average pore size of 6.33-6.58 nm, and BET surface area of 181-212 m^2/g, and the content of Al_2O_3 had significant effects on the BET surface area. The thin Al_2O_3 layer was located mostly on the surface of mesopore (as seen in X-ray photoelectron spectroscopy (XPS) spectra inFigure 4c), blocking the charge recombination during the photon-to-electron conversion process. The photoanode based on the mesoporous Al_2O_3/TiO_2 (1 mol % Al_2O_3) exhibited ~10% efficiency improvement compared to the pure mesoporous TiO_2 electrode (η increasing from 5.88% to 6.50%). Though the authors didn't compare their results with traditional nanocrystalline counterpart, the much higher surface area of the mesoporous electrode than P25 TiO_2 may possibly lead to higher dye-loading quantity, and corresponding higher light-harvesting capacity. Also, the introduction of Al_2O_3barrier layer in mesoporous TiO_2 illustrated a viable and versatile route to further improve the functions of the mesoporous photoanodes.

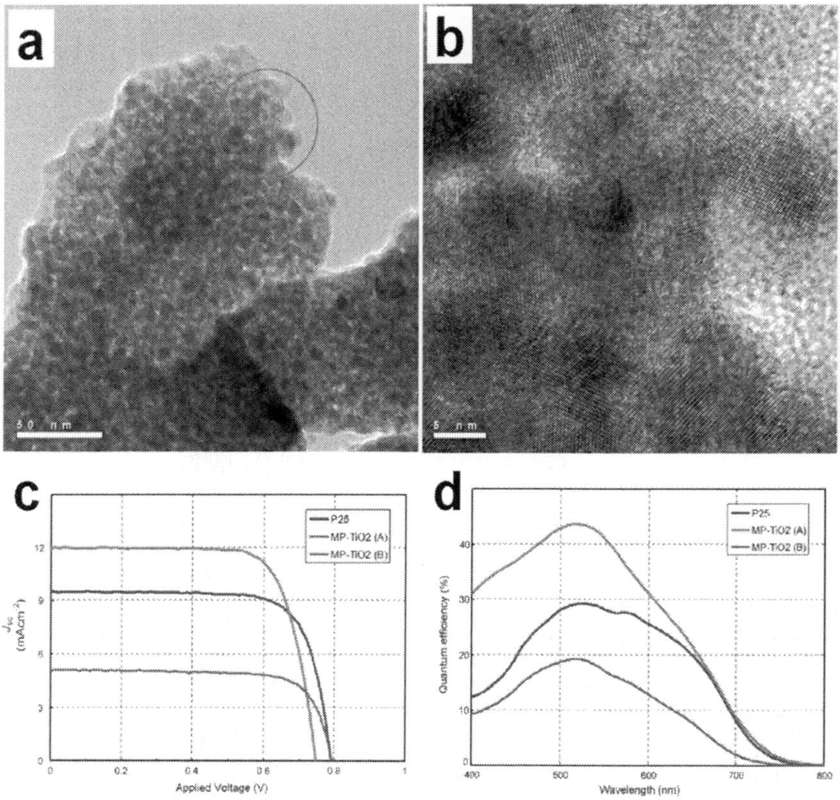

Figure 3: TEM image of mesoporous TiO_2 with a) low and b) high magnifications. c) *J*-V curves and d) IPCE spectra of DSCs prepared from P25 and mesoporous TiO_2 with $TiCl_4$ post-treatment. (Yun et al., 2011).

Though above mesoporous powders with irregular secondary particles are promising in improving the dye-loading capacity of the photoanode, it is usually difficult for them to obtain high quality electrodes with good transparency, due to the poor workability of the paste produced from the irregular micron/submicron particles. Therefore, it is intriguing to prepare mesoporous TiO_2 with regular secondary particles, e.g., TiO_2 mesoporous spheres.

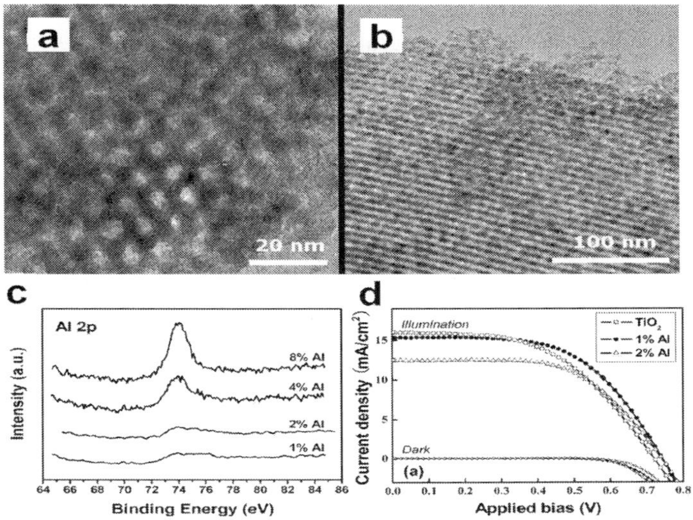

Figure 4: a-b) TEM images of mesoporous TiO_2, illustrating the ordered mesopores. c) XPS spectra of mesoporous Al_2O_3/TiO_2 with various levels of Al calcined at 525°C. d) *J*-V curves in the illuminated and dark states of pure mesoporous TiO_2, 1 and 2 mol% Al_2O_3/TiO_2 (light intensity: 100 mW/cm²; AM 1.5 filter; illumination area: 0.25 cm²). (Kim et al., 2010).

D. Chen et al. did a good job on this area, and synthesized uniform, crystalline, and mesoporous TiO_2 beads in size of ~800 nm through a combination of sol-gel and solvothermal processes, which exhibited the surface area of 108.0 m²/g and tunable pore sizes (pore diameter: 14.0 ~ 22.6 nm) (Chen et al., 2009). Due to the submicrometer-sized particle diameter and the high specific surface area, the mesoporous TiO_2 beads enhanced the LHE without sacrificing the accessible surface for dye loading, thereby increasing the PCE compared to P25 nanoparticles. The photoanode based on these mesoporous TiO_2 beads exhibited an overall light conversion efficiency of 7.20% (V_{oc} = 0.777 V, J_{sc}= 12.79 mA/cm², *FF* = 0.72), significantly higher than that derived from standard Degussa P25 TiO_2 electrode with similar thickness (5.66%).

In a further study, the group selected Ru(II)-based dye C101 and C106 to match with the mesoporous TiO_2 beads, and realized an overall PCE of greater than 10% (η=10.7%) using a single screen-printed TiO_2 layer cell construction (without an additional scattering layer).

Hence, a delicate modulation of the micron/submicron size/ shape of mesoporous materials, and a careful selection of the dye to match the dye's photon-absorption characteristics with the light scattering properties of the mesoporous materials, are important to improve the LHE and PCE of DSC (Sauvage et al., 2010).

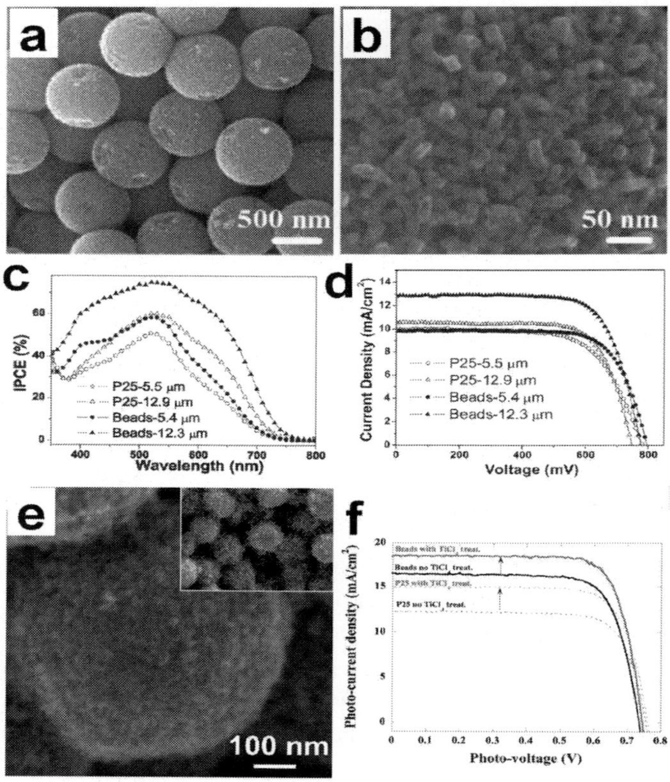

Figure 5: a-b) SEM images of the calcined mesoporous TiO_2 beads obtained after a solvothermal process with no ammonia for a) and 1.0 mL ammonia for b). c) IPCE curves and d) *J*-V curves of the TiO_2electrodes prepared from P25 NPs and mesoporous TiO_2 beads with two film thicknesses. (Chen et al., 2009) e) SEM image of the screen-printed film composed of TiO_2 porous beads; f) *J*-V curves recorded at 100 mW/cm^2 of the 12 um thick film composed of P25 particles (dashed) or TiO_2 beads (solid) sensitized with the C101 dye without (in black color) and with $TiCl_4$ post-treatment (in red color). (Sauvage et al., 2010).

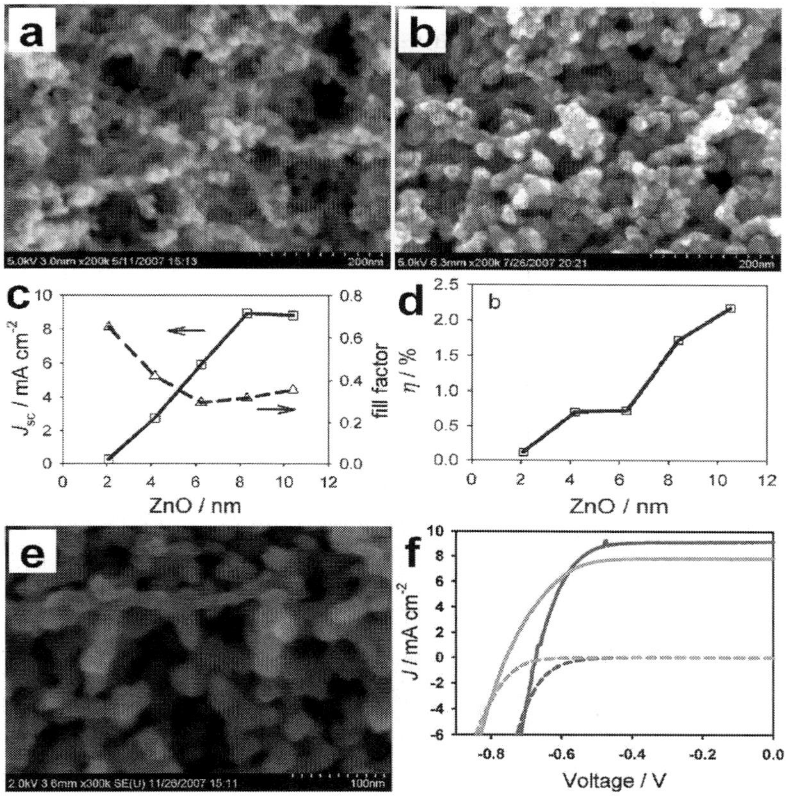

Figure 6: a-b) SEM images of aerogel frameworks coated with a) 4.4 and b) 8.4 nm ZnO. c-d) Plots of c) J_{sc}, *FF* and d) efficiency (η) as a function of thickness of ZnO deposited on aerogel frameworks. (Hamann et al., 2008a) e) SEM image of a ~30 μm thick aerogel film coated with 9.6 nm TiO_2. f) *J*-V curves for aerogel electrodes coated with 12 nm TiO_2 (Red) and TiO_2 nanoparticle (orange) in the dark (dasked lines) and under AM 1.5 illumination (solid). (Hamann et al., 2008b).

Among various mesoporous materials, aerogel is distinguished from others by its high surface area (100-1600 m^2/g) and light weight (0.003-0.5 g/cm^3) (Hüsing et al., 1998). The application of aerogel in the photoanode of DSC was pioneered by J. J. Pietron and J. T. Hamann in 2007-2008. The former work prepared the aerogel photoanode by the direct doctor-blading of the crushed TiO_2 aerogel particles, and demonstrated the IPCE of ~85% in the 500-600 nm range and ~52% at

700 nm for N719 sensitized photoanode (Pietron et al., 2007). Hamann et al. prepared SiO_2 aerogel film first via the supercritical drying process, and then coated ZnO or TiO_2 thin layer with controllable thickness via the atomic layer deposition (ALD) method (Hamann et al., 2008a, 2008b). They realized the aerogel based DSC devices with the efficiency of 2.4% (for ZnO cell) and 4.3% (for TiO_2 cell). Also, the authors investigated systematically the effects of ZnO thickness on SiO_2 aerogel template on the photovoltaic performance, and found a nearly linear relationship between the key photovoltaic index and the thickness of ZnO layer (Figure 6c,d), with the optimal ZnO thickness of 10.5 nm. These results indicated that, on one hand, the aerogel film in combination of ALD or other physical or chemical deposition processes was a powerful candidate to prepare mesoporous photoanode with controllable microstructures. On the other, the use of the dangerous supercritical drying process in the synthesis of aerogel, together with the expensive and complicated ALD process involved in the preparation of ZnO/TiO_2 layer, limited the popularization of this method greatly. There were very few studies on the aerogel photoanodes in recent five years.

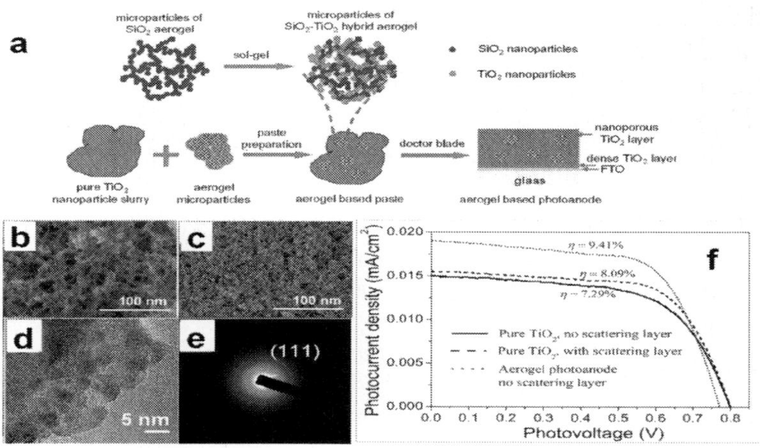

Figure 7: a) Schematic of the preparation of SiO_2-TiO_2 hybrid aerogel and aerogel based photoanode. SEM image of b) SiO_2 and c) SiO_2-TiO_2 hybrid aerogel (packing density: 0.037 and 0.125 g/cm^3). d) TEM image and e) SAED pattern of SiO_2-TiO_2 hybrid aerogel. f) *J*-V curves of DSC based on pure TiO_2 photoanode with and without scattering layer and aerogel modified photoanode. (Gao et al., 2012).

To crack above problems, X. D. Gao et al. developed a low-cost sol-gel and ambient-drying route to fabricate SiO_2-TiO_2 hybrid aerogels, which possessed a very high surface area of 500-1170 m^2/g, and incorporated them into nc-TiO_2 electrode, obtaining the aerogel hybrid photoanodes (Gao, et al., 2012). These hybrid photoanodes yielded significantly higher photocurrent densities than nc-TiO_2photoanodes due to the increased dye-loading capacity and the enhanced visible scattering effect, showing the highest efficiency of 9.41% at optimal condition, 16% higher than the $TiCl_4$-treated TiO_2photoanode modified with a conventional scattering layer. This work indicated that, the hybrid photoanodes integrating both the high-surface-area mesoporous materials and nc-TiO_2 represented a powerful and viable route toward high-efficiency DSCs.

In brief, current studies on the mesoporous photoanodes built up by the high-surface-area mesoporous materials have unambiguously proved that, the application of high-surface-area powders/beads/aerogel in the photoanode can effectively increase the dye-loaing capacity and the light scattering of the electrode, and correspondingly improve the LHE of DSC in most cases. The preparation of mesoporous materials with regular and uniform size/ shape in micron/submicron scale, and their integration with nc-TiO_2 or other nanostructures, seem feasible to achieve high conversion efficiency. A remaining problem in these high-surface-area mesoporous materials is the inferior electron-transporting property, intrinsically originating from their lower particle coordination number than the common nc-TiO_2 porous electrode. A possible solution may be the development of hybrid electrodes by incorporating metal, carbon nanotube, or graphene etc into the mesoporous structures. A mesoporous material possessing both high surface area and good conductivity is by all means necessary for highly efficient DSCs.

HIERARCHICAL PHOTOANODES

Hierarchical nanostructures, those with different morphology at different length scale and usurally characterized by the high surface area, are intresting candidates for efficienct photoanodes in DSC. This special class of nanostrcutures is spurred at the low surface area and consequently the low PCE of nanowire based photoandes, and have receivd wide-spread and intensive studies in recent years. Hierarchical ZnO and TiO_2 in the form of grass, forest, popborn ball, and tube etc. have been developed, and a wide range of chemical and physical methods have been used to prepare the hierarchical structures, including the hydrothermal synthesis, sol-gel, chemical bath deposition, pulsed laser deposition, photolithography, and so on. Due to the huge literature in this field, here we merely present some representive results in recent three years.

ZnO nanowire array may be the piorneer in the nanostructured photoanode since the outstanding work of M. Law et al. (Law et al., 2005). However, the PCE of nanowire based photoanode was low (mostly 1%-3% for a long time) mainly due to its low surface area. So the development of hierarchical ZnO and their application in DSC are wide-spread recently. As an example, Xu et al. reported a two-step synthesis process to produce hierarchical ZnO nanoarchitectures (Figure 8a,b), in which ZnO nanosheet arrays were first prepared by the pyrolysis of the precursor $Zn_5(OH)_8Cl_2$ electrodeposited on conductive glass substrates, and then dense ZnO nanowires were grown on the surfaces of the primary ZnO nanosheets by the chemical bath deposition method (Xu et al., 2010). Due to the better dye loading through the increased internal surface area and the higher light scattering behavior through extending the optical path length within the photoanode, the DSC based on the hierarchical ZnO nanowire-nanosheet architectures showed a PCE of 4.8%, nearly twice as high as that of DSC constructed using a photoanode of bare ZnO nanosheet array.

Apart from these hierarchical structures with relatively simple structures, several groups developed ZnO or TiO_2 forest (Figure 8c,d), more complicated hierarchical structures. F. Sauvage et al. prepared the hierarchical assemblies of nanocrystalline particles of anatase TiO_2 (Sauvage et al., 2010), by a pulsed laser deposition method, via the fine modulation of the plasma expansion dynamics by means of a reactive atmosphere during the ablation process. In combination with the high molar extinction coefficient heteroleptic C101 dye, the authors achieved 3.1% power conversion efficiency for 2 um thick films and 5.0% for films 7 um thick, demonstrating the great potential of the physical vapor deposition method in growing hierarchical nanostructured photoanodes with high conversion efficiency.

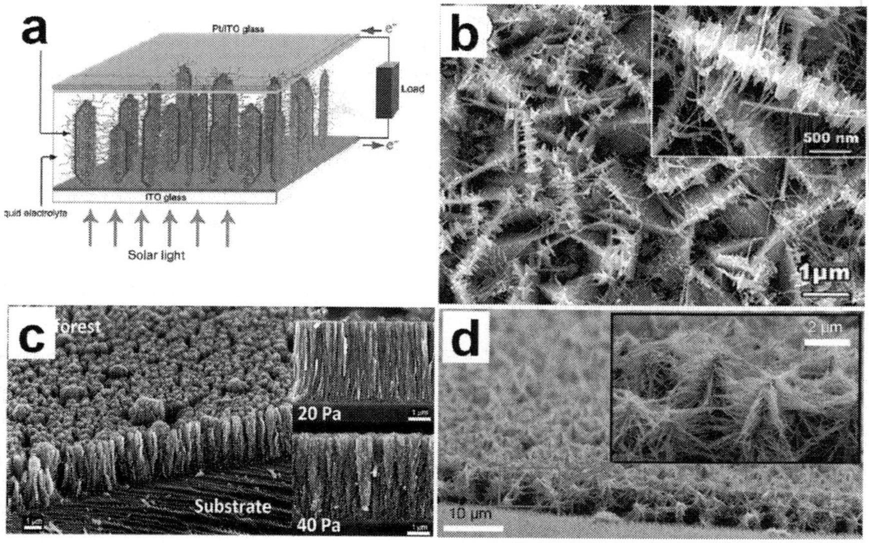

Figure 8: a) Schematic of ZnO hierarchical nanostructures derived from ZnO nanosheet arrays and b) top-view SEM image of ZnO nanowire-nanosheet architectures obtained by aqueous chemical growth after 4 h. The inset of b) corresponds to the magnified image. (Xu et al., 2010) c) TiO_2 forest-like films grown by pulsed laser deposition: (left) overview of a film deposited at 40 Pa; (right) cross sections of films deposited at 20 and 40 Pa. (Sauvage et al., 2010) d) SEM image of ZnO nanowire nanoforest (tilted view). (Ko et al., 2011).

Meanwhile, using the traditional hydrothermal method integrating with the simple selective hierarchical growth sequence, S. Ko et al. demonstrated ZnO nanoforest photoanode with high density, long branched treelike multigeneration hierarchical crystalline ZnO nanowires (Figure 8d) (Ko et al., 2011). The DSC exhibited almost 5 times higher than the efficiency of DSCs constructed by upstanding ZnO nanowires (2.63% v.s. 0.45%). The enhanced surface area for higher dye loading and light harvesting, and the reduced charge recombination by providing direct conduction pathways along the crystalline ZnO "nanotree" multi generation branches, were responsible for the efficiency improvement.

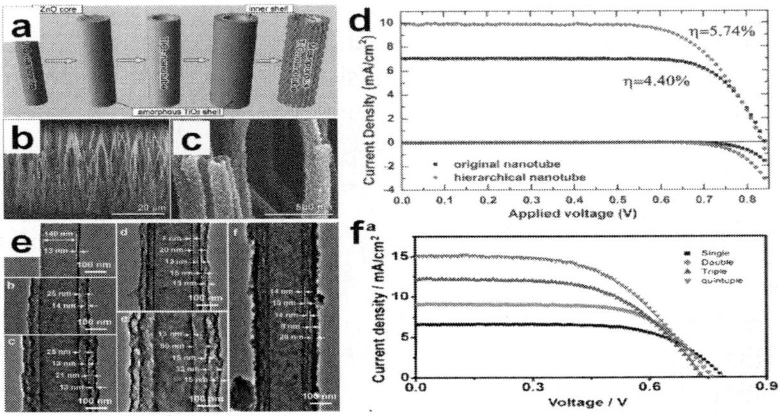

Figure 9: a) Schematic of the preparation of hierarchical TiO_2 NTA: beginning with a ZnO nanowire and followed by multiple procedures of TiO_2 coating, ZnO core removal, subsequent TiO_2 coating, and hierarchical derivation to a double-shell architecture. b) cross-section SEM image of TiO_2 nanotube array with 41 um thickness and c) magnified SEM image of hierarchical TiO_2 nanotube. d) *J–V*curves of the nanotube DSCs prepared with 20-μm thick nanotube arrays with the original and hierarchical structure. (Zhuge et al., 2011) e) TEM images of triple-shelled TiO_2 NTs with different structural features. f) *J*–V curves of DSCs based on multi-sheel TiO_2 nanotube electrodes. (Qiu et al., 2012).

Nanotube array (NTA) is another type of hierarchical structures receiving great attention, which is developed on the basis of nanowire in view of its much higher surface area than nanowire. Currently, the hierarchical TiO_2 NTA has becoming an intensive research branch. Different from the traditional synthesis of TiO_2 nanotube via the anodic oxidation, F. Zhuge et al. reported the synthesis of vertically aligned TiO_2 NTAs (up to 40 um) on FTO by using ZnO nanowire array as the hard template, and demonstrated the hierarchical derivation for effective surface area enhancement in DSCs (Figure 9 a-c) (Zhuge et a., 2011). The distinctive double-shell structure of the nanotube provided both the high surface area and the high electron-transport path, showing the overall conversion efficiency of 5.7%, 50% higher than the original nanotube (Figure 8d). In a further work, the group updated their multi-shell fabrication technology via ZnO nanowire template, and successfully obtained coaxial multi-shelled TiO_2 NTA with controllable shell numbers and shell thickness (Figure 9e) (Qiu et al.,

2012). Exhibiting the BET surface area of 119-331 m²/g, DSC devices based on these multi-shelled NTA exhibited the highest efficiency of 6.2%, higher than that based on the single-layer TiO_2 nanotube array (3.35%) at similar experimental parameters (Figure 9f).

Apart from the hierarchical nanowire/nanotube array, there are also a few studies which modulate the pores in the photoanode to realize the hierarchical structures. C. Y. Cho et al. reported an interesting method to generate hierarchical electrodes consisting of meso- and macroscale pores (Cho et al., 2011). Mesoscale colloidal particles and lithographically patterned macropores were used as the dual templates, with the colloidal particles assembled within the macropores. An infiltration of TiO_2 into the template and subsequent removal of the template produced hierarchical TiO_2 electrodes for DSC (Figure 10a). While the holographic lithography defined the macroporous networks, the colloidal crystal templates provided three-dimensionally organized mesoscale pores with a uniform size distribution (Figure 10 b-e). Owing to the strong scattering and the suppression of charge recombination in the hierarchical TiO_2 electrodes, the photovoltaic performance of the cell was comparable with nc-TiO_2 electrodes, showing a maximum efficiency of 5.0% with 50 nm pores and 6 μm thickness.

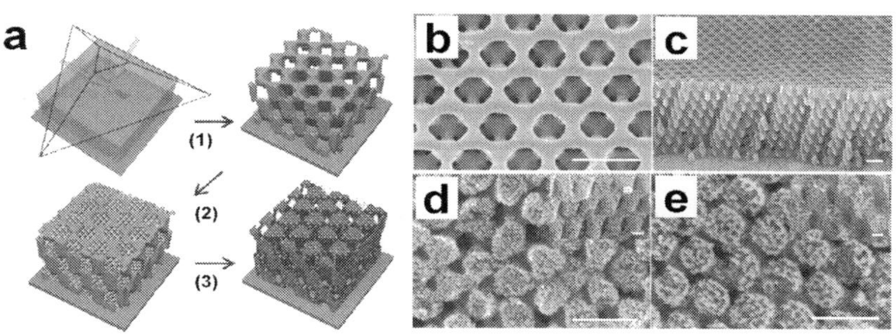

Figure 10: a) Schematic of the preparation of hierarchical TiO_2 electrode via double template method. (1) Formation of four-beam interference and the fabrication of the macroporous SU-8 structures. (2) Filling of the holographic patterns with mesoscale colloidal particles, and (3) Coating of precursors and removal of dual templates. SEM image of b) the surface and c) the cross-section of the macroporous SU-8 surface. SEM image of the hierarchical porous TiO_2 electrode from dual templates of SU-8 filled with colloidal particles with diameters of d) 60 nm and e) 110 nm. Scale bar: 1 μ m. (Cho et al., 2011).

In general, the hierarchical photoanodes, though possessing obviously higher internal surface area than the nanowire/nanotube/nanoplate counterparts and higher conductivity than nc-TiO_2 electrode, haven't show a predominant advantage over other materials up to now. The PCE of most photoanodes based on the hierarchical structures is still much lower than the state-of-the-art performance of nc-TiO_2electrode. A possible vent for the hierarchical photoanodes may be the further increase of its internal surface area while maintaining its high electron-transporting nature. Also the combination of the hierarchical structures with the mesoporous material is a meaningful and feasible strategy.

SCATTERING LAYERS ON NC-TIO_2 ELECTRODE

The photoanode of DSC is traditionally composed of nc-TiO_2 film, which is basically transparent for visible light (Rayleigh scattering, mainly), resulting in a considerable part of light shone on the DSC transmitting through the TiO_2 layer without interacting with the sensitizer. To overcome this problem, scattering layers with different structures, including scattering centers and upper scattering layers, have been employed. The theoretical studies since 1998, and subsequent substantial experimental studies on scattering layers based on the submicron particles of TiO_2, ZrO_2, ZnO, nanowire, hierarchical structures, mesoporous spheres, photonic crystals, and upper-conversion materials, comprise a vivid picture about the research activities in this area. By briefing some important advances, we describe here some novel scattering layers developed in recent years.

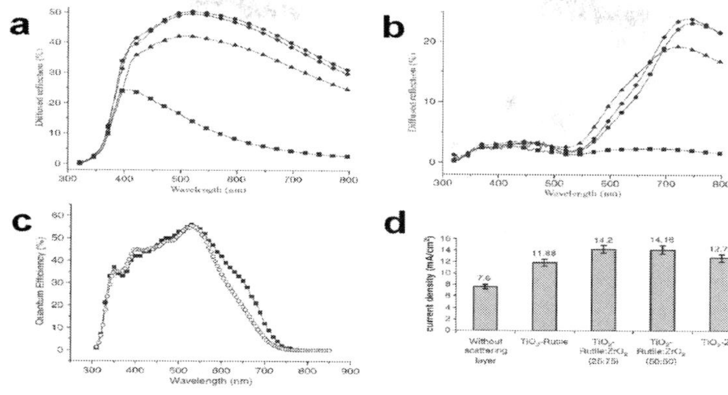

Figure 11: a-b) Diffused reflection spectra of a) undyed and b) sensitized TiO_2 layer without scattering layer (■), with the scattering layers of TiO_2-Rutile (▲), ZrO_2 (●) and 25:75 mixture of TiO_2-Rutile and ZrO_2 (◆). c) IPCE curves for devices with TiO_2 electrode with (■) and without (○) a scattering layer, the scattering layer being 25:75 mixture of TiO_2-Rutile and ZrO_2. d) Current densities obtained from devices with 4 mm thin layer of TiO_2 with and without scattering layers. The thickness of the scattering layers is ~6 um for all samples. (Hore et al., 2006).

In 1998, J. Ferber et al. started from Mie theory, and calculated the multiple scattering in TiO_2electrodes using a numerical solution of the radiative transfer equation (Ferber et al., 1998). These calculations predicted that, a suitable mixture of small particles (20 nm), which resulted in a large effective surface, and of large particles (250-300 nm), which were effective light scatterers, had the potential to enhance the solar absorption significantly.

In 2006, S. Hore et al. reported systematically the effects of TiO_2/ZrO_2 scattering layer on the photovoltaic performances (Hore et al., 2006). Different scattering layers composed of commercial TiO_2-Rutile (Bayer Germany) and ZrO_2 (TOSOH Corporation, Tokyo, Japan) with varied proportion were examined, providing an in-depth understanding on the functions of the top scattering layer. DSC device with an area of 2.5 cm^2 exhibited the PCE of 6.8% using an only 4 μm layer of TiO_2 electrode together with an optimal scattering layer, illustrating the potential of a suitable scattering layer on both the improvement of the efficiency and the reduction of the cost by using less sensitizer.

Very recently, F. E. Gálvez et al. reported an integral optical and electrical theoretical analysis on the effects of the different design of the diffuse light scattering on the performance of DSCs (Gálvez et al., 2012). Based on a Monte Carlo approach, they introduced the light harvesting efficiency and the electron generation function extracted from optical numerical calculations in a standard electron diffusion model, to obtain the steady-state characteristics of the different configurations considered (Figure 12). They proposed that the diffuse scattering layers acting as the back reflector provided the largest achievable light harvesting efficiencies, which determined an optimum overall performance, for electron diffusion lengths longer than the electrode thickness. When the electron diffusion length was shorter than the electrode thickness, the embedding of the diffuse scattering particles in the nanocrystalline paste yielded a better output even when the light harvesting was not optimal. These results can provide helpful guidance for us to design and prepare the scattering structure of the photoanode.

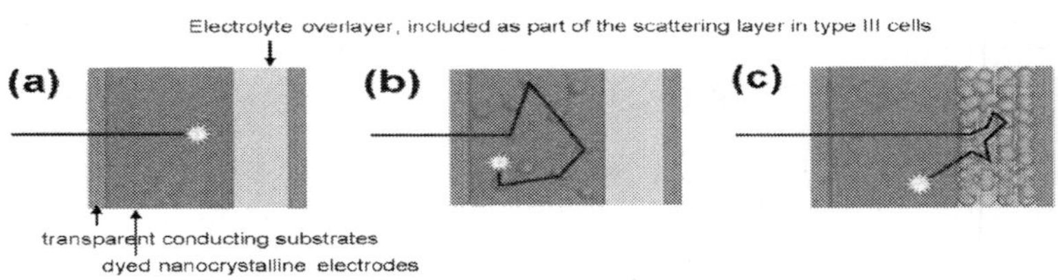

Figure 12: Simulated trajectories of a photon absorbed by a) standard semi-transparent cell, b) cell with an electrode embedding diffuse scattering particles, and c) cell made of a semi-transparent electrode coupled to a diffuse scattering layer, illustrating the different series of scattering events that yield longer optical paths and thus enhance the probability of absorption in each type of modified cell under consideration. (Gálvez et al., 2012).

In recent years, with the rapid development of nanostructured photoanodes, various nanostructures have been used as the top scattering layer of nc-TiO_2 electrode or the novel photoanodes based on nanowire/ nanotube arrays etc. For example, K Fan et al. applied TiO_2 fusiform nanorods (diameter: 20-80 nm, length: 200-400 nm) as the scattering layer of P25-based electrode (Figure 13), and observed 66.5% improvement in the efficiency with lower resistance and longer electron lifetimes as compared to the bare P25-based solar cell (Fan et al., 2011). The reduced charge recombination and the sufficient scattering effect of the nanorods in the film electrode were responsible for this improvement.

Koo H. J. et al. prepared nano-embossed hollow spherical (NeHS) TiO_2 particles and investigated their photovoltaic property in DSC as the bifunctional scattering layer (Koo et al., 2008). The walls of the obtained hollow spheres were composed of nanocrystalline anatase TiO_2 particles with mesoporous structure (Figure 14 a-c), which endowed them 5 times higher dye-loading capacity (N719 dye: up to 5.0×10^{-5} mol/g) than that of the normally used 400 nm-diameter scattering particles with flat surfaces. The NeHS particles scattered incoming light effectively as confirmed by the reflectance spectroscopy (Figure 14 d,e). When the NeHS TiO_2 particles were used as the secondary scattering layer in DSC, a substantial improvement in the efficiency has been achieved, with the highest conversion efficiency of 10.34%. Huang F et al. used the submicrometer-sized mesoporous TiO_2 beads as the dual function scattering layer for the nanocrystalline electrode, and observed similar results, demonstrating that the dual-function scattering layer represented a powerful alternative for the traditional large-particle scattering layer for high-efficiency DSCs (Huang et al., 2010).

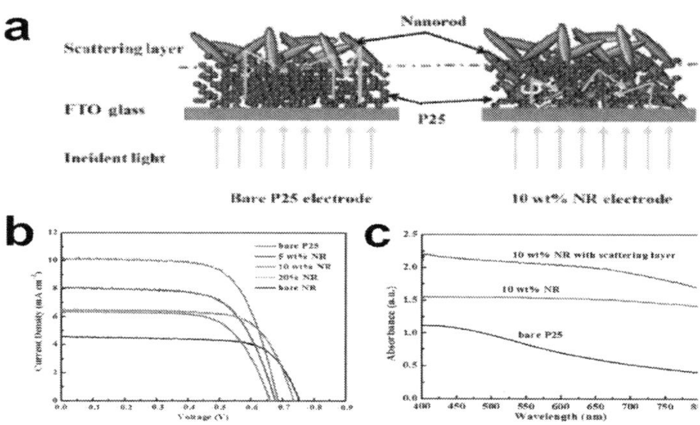

Figure 13: a) Schematic of the photoanodes with nanorod scattering layer. b) *J*-V curves of DSCs, and c) UV-VIS spectra of bare P25 film, and with different nanorod scattering layer. (Fan et al., 2011).

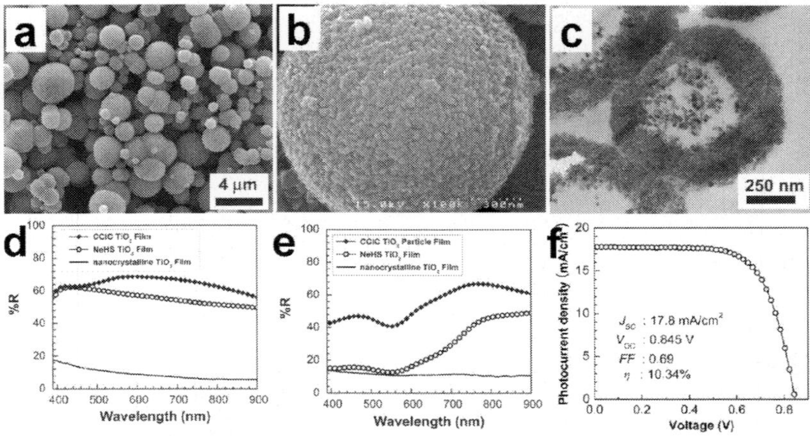

Figure 14: SEM image of a) the as-synthesized nano-emboded hollow sphere (NeHS) TiO_2 particles, b) high-magnification image for NeHS TiO_2 calcined at 450 °C. c) TEM image of a sliced NeHS TiO_2particle. d-e) Diffused reflectance spectra of the nanocrystalline, CCIC, and NeHS TiO_2 particulate films d) without and e) with adsorbed N-719 dye. f) *J*-V curve of a DSC based on NeSH TiO_2particulate film as an overlayer on a nanocrystalline TiO_2 film under AM 1.5G-one sun light intensity. (Koo et al., 2008).

Below we gave two examples which used the nanostructured scattering layer on the electrodes based on nanowire array or nanofibers. Shao et al. demonstrated an interesting double layer photoanode, with the bottom layer of TiO_2 nanorod array providing direct conduction pathway for photo-generated electrons, and the upper layer of microflowers built up by TiO_2 nanobelt, increasing the light harvesting ability as the scattering part (Shao et al., 2011). The cell based on this hierarchical anatase TiO_2 exhibited a conversion efficiency of 5.53%, superior than commercial TiO_2 (P25). The much higher optical reflectance of the hierarchical structures than the nanorod array and nanoparticle films (Figure 15) proved that this morphology was beneficial to the light-scattering capacity of the photoanode.

Yang et al. reported an innovative bilayer TiO_2 nanofiber photoanode, which combined both smaller (60 nm) and larger (100 nm) diameter TiO_2 nanofibers fabricated by electrospinning (Yang et al., 2011). The smaller diameter nanofiber (SNF) layer with a high surface-to-volume ratio was used to adsorb sufficiently dye molecules and directly transport electrons released from excited dyes. The bigger-diameter nanofi ber (BNF) layer worked as light scattering, adsorbed sufficient dye molecules for energy harvest, and provided higher pore volume in BNF to facilitate electrolyte diffusion for regenerating sensitized dye molecules in the photoanode. Therefore, this bilayer composite nanostructure photoanode offered excellent dye-loading, lightharvesting, and electron-transport properties. The PCE of DSC was improved from 7.14% for the single-layer to 8.40% for the bilayer of TiO_2 nanofiber photoanode, representing an increase of 17%.

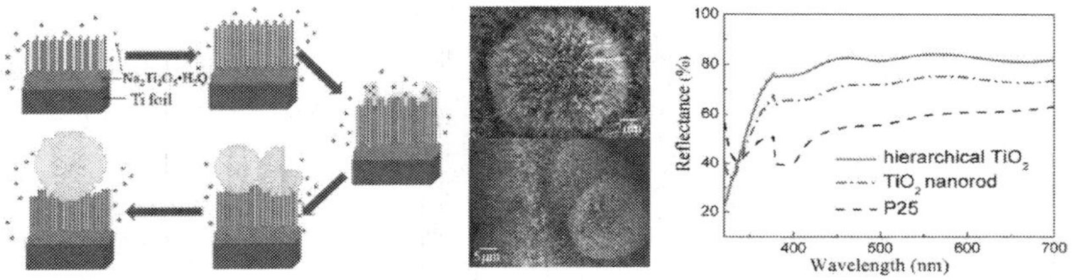

Figure 15: Schematic of TiO_2 nanorod array-micro flower hierarchical photo-anode, SEM images of the photoanode at surface and cross section, and reflectance spectra of three kinds of TiO_2 photoanode on FTO substrate. (Shao et al., 2011).

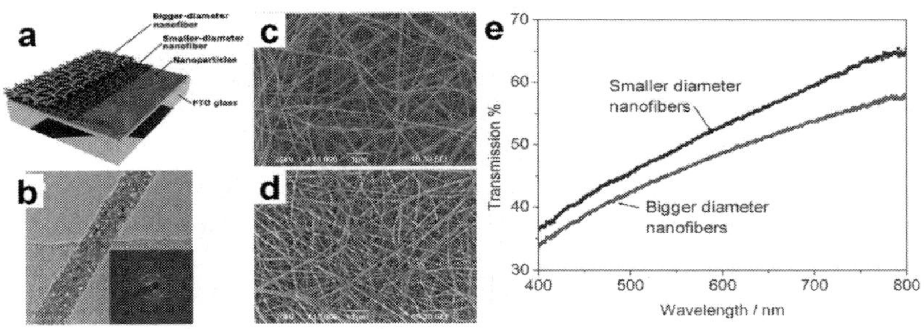

Figure 16: a) Schematic of the photoanode based on bi-layer electronspinned TiO_2 nanofibers. b) TEM image of smaller nanofiber and corresponding SAED pattern. c-d) SEM images of c) smaller and d) bigger nanofibers. e) UV-VIS transmission spectra of smaller diameter and bigger-diameter nanofiber photoanode with the same thickness (Yang et al., 2011).

As the conclusion of this section, we introduce a special scattering layer using up-converting materials. G. B. Shan et al. utilized the hexagonal nanoplatelets of β-$NaYF_4$:Er^{3+}/Yb^{3+}, a typical up-conversion phosphor owing to Er^{3+} doping with Yb^{3+} codopant assisting in the energy transfer process, as the second layer on nc-TiO_2 electrode (Figure 17) (Shan et al., 2011). Approximately 10% enhancements of photocurrent and overall DSC efficiency were achieved by the addition of the external layer, which exhibits two functions of light reflecting and near-infrared (NIR) light harvesting. Though the overall PCE of the cell and the improvement were moderate, the work opened a new opportunity to use some advanced functional materials to enhance the light-harvesting functions of DSC.

In summary, a wide variety of nanostructures including submicron particles, nanowires, nanofibers, mesoporous beads, spheres, nanopalletes etc have been used as the top scattering layer of DSC, which are proved to be very effective in improving the LHE and PCE of DSC, especially for the dual-function materials possessing both high surface area and large-particle size. At the same time, the application of multiple function materials such as the up-converting nanopalletes in the scattering layer sheds a new light for further development.

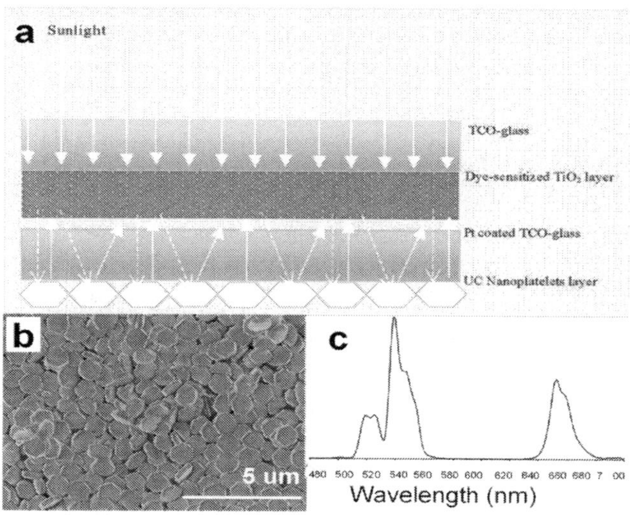

Figure 17: a) Schematic of the DSC device consisting of one internal TiO_2 transparent layer plus an external rear layer of β-$NaYF_4$:Er^{3+}/Yb^{3+} nano-platelets. b) SEM image and c) up-conversion fluorescence spectrum of the β-$NaYF_4$:Er^{3+}/Yb^{3+} nanoplatelets. (Shan et al., 2011).

PLASMONIC DSCS

Localized surface plasmon resonance (LSPR) behavior is an intriguing characteristic of metal nanoparticles (NPs), which is generated by the resonance between electric fields of electromagnetic waves and free electrons in metal NPs. The use of plasmonic effects has been proposed as a promising pathway to increase the light absorption in active layers of solar cells, and has been demonstrated on several thin-film solar-cell materials such as amorphous silicon, gallium arsenide, polymers, and DSC. Previous studies on the plasmonic solar cells with thin active layers have proved the beneficial functions of the LSPR on the photocurrent density through such following channels:

1. excitation of localized surface plasmon resonances of metallic NPs;
2. scattering of light by metallic NPs into dielectric-like waveguide modes of the solar cell;
3. coupling to propagating surface plasmon polariton (SPP) modes (Ding et al., 2011).

This has established a solid foundation for the rapid development of plasmonic DSCs in recent years.

The mostly-used metal NPs in the plasmonic DSCs are Ag and Au. While earlier studies have found the corrosion of metal particle in the electrolyte or the undesired introduction of electron-hole recombination by the plasmonic particles, it has become a common practice to protect the Ag or Au NPs by a thin TiO_2 or SiO_2 layer. Coupled with the standard thick nc-TiO_2 electrode, these plasmonic photoanodes performed fairly well compared with the control samples.

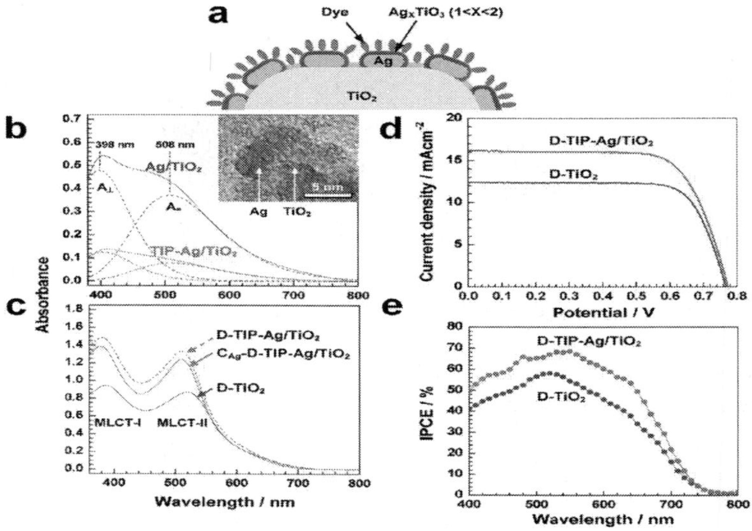

Figure 18: a) Schematic of Ag NPs deposited on TiO_2 NPs. b) Plasmonic absorption spectra of Ag NPs in Ag/TiO_2 (blue solid curve) and TIP-Ag/TiO_2 (red solid curve). Each spectrum is decomposed into two absorption peaks depending upon the geometry of Ag NPs. The inset shows a TEM image of the side view of Ag NP on TiO_2. $Ag_{\perp}$ and $Ag_{=}$ indicate the direction in the geometry of Ag NP. c) UV-VIS absorption spectra of D-TiO_2, D-TIP-Ag/TiO_2, and C_{Ag}-D-TIP-Ag/TiO_2. d) *J*-V and e) IPCE curves of D-TiO_2 and D-TIP-Ag/TiO_2 photoanodes. TIP-Ag: TiO_2 coated Ag NPs; D-TiO_2: N719 dye sensitized TiO_2 electrode; C_{Ag}-D-TIP-Ag/TiO_2: the extinction spectrum after removing the contributions of Ag NPs from the spectrum of D-TIP-Ag/TiO_2. (Jeong et al., 2011).

N. C. Jeong et al. provided a detail examination on the effects of the protection of Ag NPs on the LSPR and the cell performance (Jeong et al., 2011). They deposited Ag NPs on TiO_2 framework via the photo-reduction of Ag^+ from dissolved $AgNO_3$, and protected them by a thin TiO_2 layer by refluxing Ag-TiO_2 electrode in isopropyl alcohol and subsequent annealing at 370°C. The electrode incorporating the protected silver NPs (Ag-TiO_2) showed enhanced extinction of a subsequently adsorbed dye (the ruthenium-containing molecule, N719), realizing an overall conversion efficiency of 8.9% and 25% improvement over the performance of otherwise identical solar cells incorporating silver NPs lacking protection (Figure 18). Roughly half the improvement could be traced to the increased dye loading by the photoanodes following silver incorporation, with the remaining improvement coming from the localized surface plasmon resonance (LSPR) enhancement of the effective absorption of N719 dye molecules.

J. Qi et al. coated TiO_2 thin layer (~2 nm) on Ag NPs and incorporated this Ag@TiO_2 nanostructures into the TiO_2 photoanode (Qi et al, 2011), which can easily transferred the carriers to surrounding TiO_2 NPs in contact with the shell, while providing a good protection for the metal NPs. By utilizing Ag@TiO_2 core-shell nanostructures, the optical absorption of dye molecules in solution and in thin film was enhanced by the strong localized electric field generated by LSPs (Figire 19 a,b). By incorporating Ag@TiO_2 NPs, the PCE of DSCs with very thin photoanodes (1.5 μm) was increased from 3.1%to 4.4%, and a small amount of Ag@TiO_2 NPs (0.1 wt%) improved efficiency from 7.8% to 9.0% while decreasing the photoanode thickness by 25% for improved electron collection (Figure 19c,d).

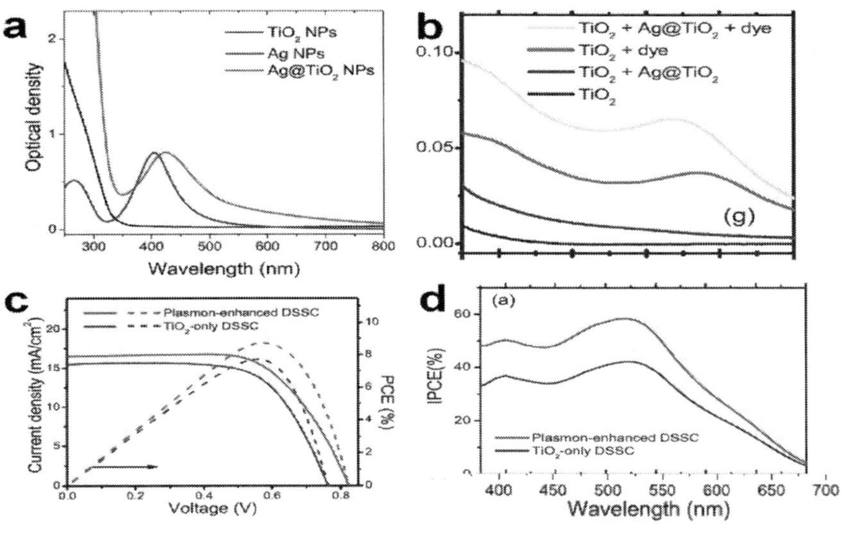

Figure 19: a) Absorption spectra of Ag NPs solutions stabilized by PVP, TiO_2 NPs, and Ag@TiO_2 NPs. b) Absorption spectra of Ag@TiO_2 NPs, dye, and their mixtures in TiO_2 film. c) *J*-V curves of plasmonic DSC (Ag/TiO_2 = 0.1 wt %, $\eta = 9.0\%$, $FF = 67\%$, 15 μm) and TiO_2-only DSC ($\eta = 7.8\%$, $FF = 66\%$, 20 μm). d) IPCE spectra of the DSCs with and without Ag@TiO_2. (Qi et al, 2011).

M. D. Brown et al. integrated Au NPs, which exhibited intense absorption due to the surface plasmon resonance in the visible band, into both the liquid and solid state DSCs (Brown et al., 2011). They successfully overcame the detrimental effects of the incorporating "bare" metal NPs into the bulk DSCs, by coating the Au NPs with a thin shell of silica, which could sustain the sintering process of the TiO_2 with negligible influence to the optical properties, resist the corrosion from the iodide/triiodie electrolyte, and enhance the photocurrent generation and solar cell performance as a result of light harvesting by the metal NPs. With a significant increase in the short circuit photocurrent, the best solid state plasmonic cell sensitized with Z709 exhibited a PCE of 4.0% (Figure 20). The readily tunable optical properties of metallic NPs, through the use of structures such as nanobars, nanostars, or other complex shapes, provided extensive scope for further work to fully optimize the technology.

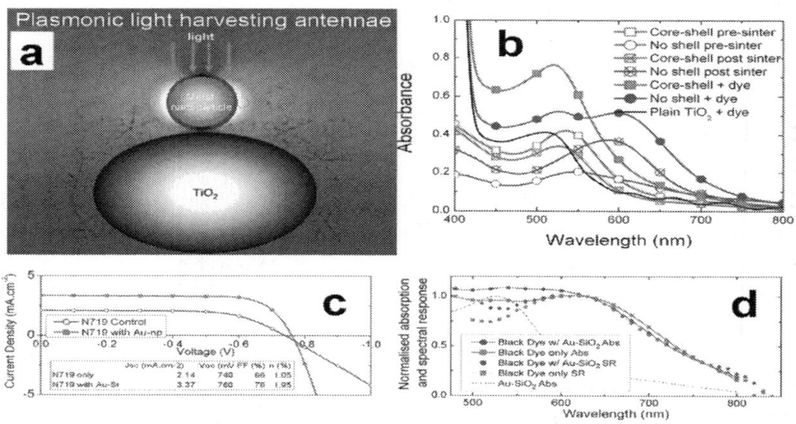

Figure 20: a) Schematic of plasmonic Au-TiO_2 structures. b) Absorption spectra of thin films (~1 μm) processed with nc-TiO_2 paste at different fabrication stages, including directly after doctor blade coating (pre-sinter), after two sintering cycles at 500°C and a $TiCl_4$ treatment (post-sinter), with final sensitization of Z907 (dye), TiO_2 paste incorporating bare Au NPs, Au-SiO_2 core shell NPs, and nc-TiO_2 film sensitized with Z907. c) *J*-V curves of liquid DSC (N719 sensitization, 1.1 μm thickness), with and without Au-SiO_2 core (15 nm)-shell (3 nm) nanoparticles. d) Normalized absorption (as 1-transmission-reflection) and spectral response of liquid DSCs. (Brown et al., 2011).

H. Choi et al. investigated the effects of the oxide capping layer of Au nanoparticles on the performance of plasmonic DSCs (Choi et al., 2012). By employing SiO_2- and TiO_2-capped Au NPs, they improved the cell efficiency from 9.3% for an N719 sensitized device to 10.2% upon incorporation of 0.7% Au@SiO_2 and to 9.8% upon loading of 0.7% Au@TiO_2 NPs. The plasmonic effect was observed in Au@SiO_2 incorporated DSC, which produced higher photocurrent (Figure 21). However, in Au@TiO_2 incorporated DSC, Au NPs underwent charge equilibration with TiO_2 NPs and shifted the apparent Fermi level of the composite to more negative potentials, which resulted in a higher photovoltage. These observations opened up new opportunities to introduce both these paradigms and to synergetically enhance the photocurrent and photovoltage of DSC.

In brief, the plasmonic photoanodes are still a newly-emerged research area in DSC, spanning a wide range of materials (metal, semiconductor) and optical and electrochemical phenomena (localized surface plasmon resonance, scattering, electron-hole recombination). It is currently the main task for scientists to seek a suitable plasmonic structures (metal-oxide core-shell structures) and effective strategies to incorporate them into the nanocrystalline or nanostructured electrodes. While at present the nanoparticles of Au and Ag have been attempted to produce plasmonic DSC, further efforts may be extended to other metals (Ti, Zn, etc.) or morphologies with complex structures of metals (cubic, hexagonal, star-form etc.). Also, in-depth explorations on the mechanism of the plasmonic resonance embedded in the nanoporous electrode, and their detail influences on the photo-to-electric conversion process, are very useful to promote the development of this intriguing area.

a

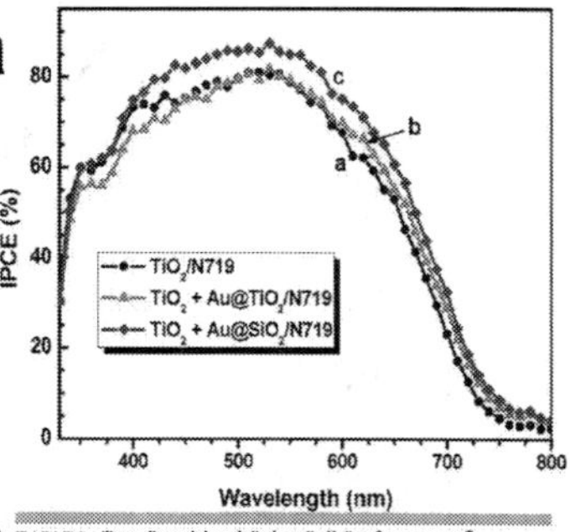

b

TABLE 1. Dye-Sensitized Solar Cell Performance[a]

support/dye	J_{sc} (mA cm^{-2})	V_{oc} (V)	ff	η (%)
TiO_2/N719	18.28	0.729	0.697	9.29
TiO_2+Au@TiO_2/N719	18.281	0.771	0.694	9.78
TiO_2+Au@SiO_2/N719	20.31	0.727	0.691	10.21

[a] Performances of DSSCs were measured with 0.18 cm^2 working area under AM 1.5 illumination. Electrolyte: 0.6 M DMPImI, 0.05 M I_2, 0.1 M LiI, and 0.5 M *tert*-butylpyridine in acetonitrile. Au@TiO_2 and Au@SiO_2 loadings were kept at 0.7% by weight. ff and η correspond to fill factor and power conversion efficiency, respectively. DMPImI, 0.05 M I_2, 0.1 M LiI, and 0.5 M *tert*-butylpyridine in acetonitrile. Excitation: AM 1.5 white light.

c

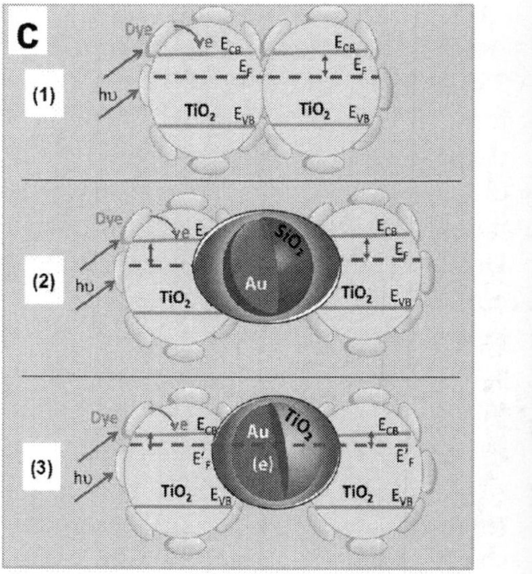

Figure 21: a) IPCE spectra of DSC employing N719 adsorbed onto TiO_2, TiO_2/Au@TiO_2, and TiO_2/Au@SiO_2as photoanodes. The loading of core-shell particles is maintained at 0.7%. b) Photovoltaic performance of DSCs. c) Electron equilibration and its influence on the apparent Fermi level (EF): (1) dye TiO_2, (2) dye TiO_2/Au@-SiO_2, and (3) dye TiO_2/Au@TiO_2. LSP influence is seen in both (2) and (3), and shift in Fermi level as a result of electron accumulation in the metal core is seen in only (3). (Choi et al., 2012).

PHOTONIC CRYSTALS BASED PHOTOANODES AND MORE

Photonic crystals (PCs) are materials that exhibit periodicities in their refractive index on the order of the wavelength of light, and thus provide many interesting possibilities for "photon management" (John, 1987). PC layers can be coupled to the charge-generating layer of DSC, and increase the LHE via such following mechanisms,

1. photon localization and enhanced red light absorption near the edges of the photonic bandgap,
2. light reflection within the photonic bandgap at various angles, and

3. formation of photon resonance modes within the dye-sensitized layer (Yip et al., 2010).

Compared with the traditional scattering techniques using geometrical optic-based elements (incoherent scattering layers, or high-reflectivity metallic mirrors), the application of PCs can obtain a more significant improvement on the LHE via the delicate control of the reflected, diffracted, or refracted light passing through the cell, apart from the apparent advantage in obtaining the transparent and colorful DSC rather than the opaque ones. The major limitation for the incorporation of PCs (especially the self-assembled 3D PCs) into DSC is the incompatibility between the fabrication routes for the photonic structures and the photoanode. So in the current stage, the preparation technology of PCs and their coupling to nc-TiO_2 layer represent the emphasis of many studies. In this section, we give several examples of PC layer in DSC with distinctive features in their structure.

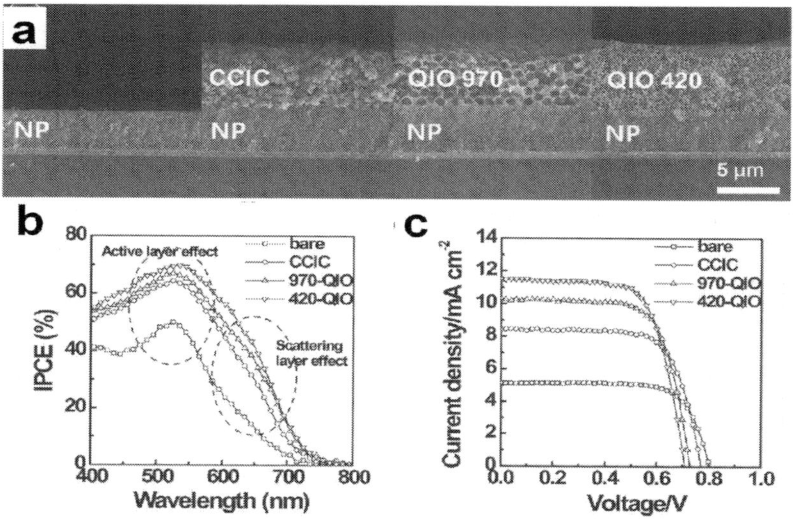

Figure 22: a) Cross-sectional SEM images of the photoelectrodes of nc-TiO_2 and with different scattering layers. b) IPCE spectra and c) *J–V* curves of the DSCs employing different scattering layers. (Han et al., 2011).

S. Han et al. fabricated quasi-inverse opal (QIO) layers with imperfect periodicity of hollows and applied them as the new scattering layer in DSC (Han et al., 2011). The porous QIO layers composed of highly crystalline anatase TiO_2 nanocrystals exhibited good dye-adsorptive properties and effective light-scattering properties over the wavelength range of 600–750 nm. The photocurrent of DSC based on the 420 nm-QIO layer was higher than the commercial scattering layer-based DSSC, achieving an efficiency of 5.7% (Figure 22).

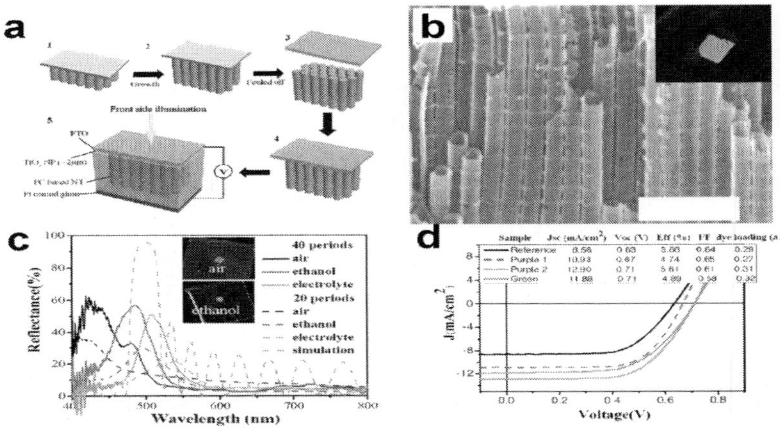

Figure 23: a) Schematic of the cell fabrication process. 1) Formation of PC layer. 2) Formation of NT layer. 3) Detachment of bi-layer from substrate. 4) Gluing to FTO by TiO_2 NPs. 5) Assembly of DSC. b) Cross-section SEM image of the PC layer. Scale bar:1 μ m. c) Reflectance of PC layers with 20 and 40 periods (lattice constant ~ 150 nm). Insets: photographs of the samples (with 20 periods) in air and infiltrated with ethanol. The green curves are simulated results with lattice constants of 150 and 190 nm for purple and green samples. d) *J*-V curves under AM 1.5 solar light illumination. The axial lattice parameters in the PC layer of the purple and green cells are ~ 150 and ~ 190 nm, and the numbers of periods are 30 and 20, respectively. Purple 1 and purple 2 are similar cells with a slight difference in dye loading. (Yip et al., 2012).

To solve a series of problems related to the PC coupled DSCs, including the poor physical contact between the PC layer and the TiO_2 absorbing layer, and the poor charge transport due to the nonconductive nature of PC, etc., C. T. Yip et al. developed a seamless PC-TiO_2 nanotube assembly (Yip et al., 2012), where the TiO_2 nanotube (NT) layer was fabricated by normal electrochemical anodization and the TiO_2 PC layer was obtained by a periodic current pulse anodization (Figure

23). Corresponding DSCs showed a 50% efficiency improvement compared with the cells without a PC layer, due to the enhanced light harvesting of the DSCs in the spectral range corresponding to the photonic bandgap of the PC and a longer wavelength range.

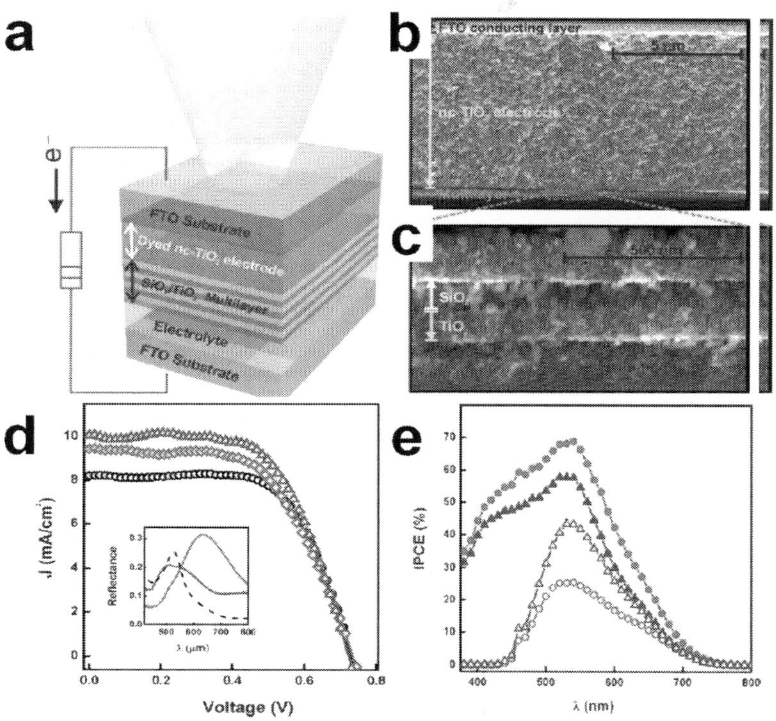

Figure 24: a) Scheme of DSC based on the 1D PC exposed to frontal illumination. b) Cross-section SEM image of nc-TiO_2 electrode, onto which a TiO_2–SiO_2 NP multilayer is deposited. c) Magnified view of the SiO_2 and TiO_2 layers composing the 1D PC. d) *J*–V curves of nc-TiO_2 electrode (7.5 mm thick) coupled to different 1D PCs under one-sun illumination. The lattice parameters are 120±10 nm (open green triangles), and 160±10nm (open red diamonds). Reference cell with the same nc-TiO_2electrode is plotted as the open black circle line. Inset: reflectance spectra of the PC-based cell together with the absorption spectrum of the ruthenium dye. e) IPCE spectra for a DSC containing a 7.5 mm thick dye-sensitized TiO_2 electrode (solid blue triangles) and for the same electrode coupled to a PC of parameters 120±10nm (solid red circles), measured under frontal illumination and rear illumination (open blue triangles and open red circles, respectively). (Colodrero et al., 2009).

S. Colodrero, et al. couped a porous and highly reflecting 1D photonic crystal into a nc-TiO_2electrode, by the deposition of alternate layers of NPs of different compositions by spin-coating (Colodrero et al., 2009). While the porous mesostructure allowed the electrolyte to flow through it and soak the electrode without interfering with the charge transport through the cell, the PC layer with a thickness of just half a micrometer efficiently localized the incident light within the nc-dyed TiO_2electrode in a targeted wavelength range (Figure 24). Consequently, the average PCEs were improved to between 15 and 30% of the reference value attained for standard electrodes, while the open-circuit voltage and the transparency of the cell remained intact, contrary to what happened when the scattering layers were employed to improve the light harvesting.

Apart from the intensive studies on the PCs based DSC, other optical elements such as planar waveguide were also explored to enhance the LHE of DSC. Typically, Z. L. Wang's group demonstrated a novel three-dimensional DSC by integrating planar optical waveguide and nanowires (NWs) (Wei et al., 2010). The ZnO NWs were grown normally to the quartz slide, and the 3D cell was constructed by alternatively stacking a slide and a planar electrode. While the slide served as a planar waveguide for light propagation, the 3D structure effectively increased the light absorbing surface area due to internal multiple reflections without increasing the electron path length to the collecting electrode. Obtained 3D DSCs exhibited a significant improvement in energy conversion efficiency by a factor of 5.8 compared to the planar illumination case.

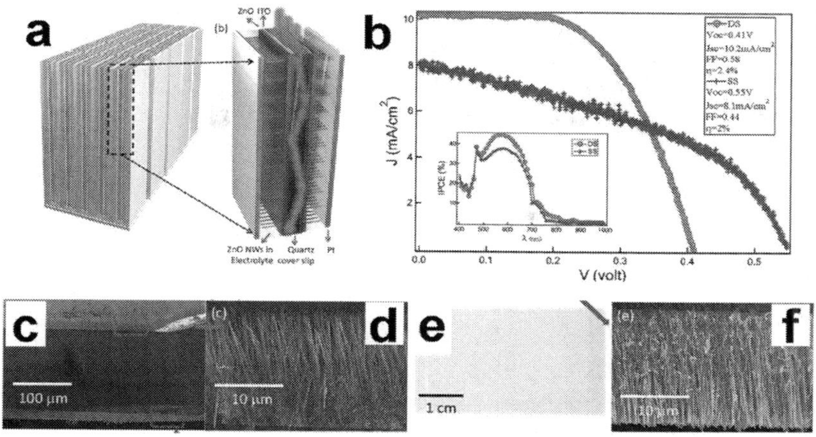

Figure 25: a) Schematic architecture of large scale 3D DSSC. The waveguide-NW 3D unit cells areplugged into the counter electrode housing and then sealed and fully packaged. b) *J*-V curves of DSCs under one full sun illumination for the parallell to the waveguide surface (PS) configuration. Inset: typical IPCE curves measured for single-side (SS) and double-side (DS) coated DSCs in the PS case. c) Low-magnification SEM image of a quartz slide with uniformly grown ZnO NWs on DS surfaces. d, f) High-magnification SEM images showing the densely packed ZnO NWs on top and bottom surfaces of the slide, respectively. e) Image of a slide coated with grown ZnO nanowire arrays. (Wei et al., 2010).

In summary, the optical elements such as the photonic crystals and the planar waveguide can provide exciting perspective to deliberately modulate the optical path of the incident lights and to significantly enhance the LHE of DSC. However, the ultimate PCE of corresponding DSCs is still far lower than that of nc-TiO_2 counterpart. How to tackle with the preparation problems related to the PC layers, and to avoid the possible adverse effects of the PC layers on the dye-loading and charge-transporting properties of the original electrode, are the kernel issues which should be solved in near future.

CONCLUSIONS

In conclusion, the light-harvesting efficiency (LHE) of the photoanode film has determinative effects on the power conversion efficiency of DSC.

The deliberate modulations of the internal surface area of the nanoporous electrode and the optical path of the incident light are currently the main pathway to enhance the LHE of DSC. A wide range of novel materials or techniques have been utilized to improve the LHE of the electrode, including the high-surface area mesoporous nanostructures or aerogel, scattering-enhanced hierarchical nanostructures, up-conversion materials, plasmonic core-shell structures, and photonic crystals etc. However, while most of reported work realized obvious enhancement on one or more specific capacities of DSC, such as the dye-loading properties, optical scattering, or improved harvesting of near-infrared light, very few study can demonstrate high device performance comparable with the state-of-the-art nc-TiO_2 cell. The intrinsically different particle size, microstructures, preparation strategy of these novel materials from the traditional nc-TiO_2 electrode will inevitably result in significant changes in the microstructure or the optical/electrical properties of the photoanode, which may greatly impair the final performance of the device. How to balance the advantageous and disadvantageous factors involved in these new-type photoanodes, and realize the solid improvement of the overall performance of DSC, are the emphasis of the scientists in near future. After all, the photoanodes based these novel materials or structures are still in an infant stage, containing infinite possibilities to improve or even revolutionize the basic principle and performance of the traditional DSC. We believe, via the intensive and extensive efforts of the scientist all over the world and through the collaborative studies among different areas spanning from material, chemistry to physics, the new-type photoanodes will certainly usher a brilliant future.

ACKNOWLEDGEMENTS

This work is supported by the 973-project (Grant no. 2009CB623304) of Ministry of Science and Technology of China and the Basic Research Program (Grant no. 51072214, 51002174) of National Natural Science Foundation of China.

REFERENCES

1. M. D. Brown, T. Suteewong, R. S. S. Kumar, V. D. Innocenzo, A. Petrozza, M. M. Lee, U. Wiesner, H. J. Snaith, 2011 Plasmonic Dye-Sensitized Solar Cells Using Core−Shell Metal−Insulator Nanoparticles Nano Lett. 11 2 438 445
2. W. Chen, X. Sun, Q. Cai, D. Weng, H. Li, 2007 Facile Synthesis of Thick Ordered Mesoporous TiO2 Film for Dye-Sensitized Solar Cell Use Electrochem. Comm. 9 3 382 385
3. D Chen, F Huang, Y. B. Cheng, R. A. Caruso, 2009 Mesoporous Anatase TiO2 Beads with High Surface Areas and Controllable Pore Sizes: A Superior Candidate for High-Performance Dye-Sensitized Solar Cells Adv. Mater. 21 20 2206 2210
4. C. Y. Cho, J. H. Moon, 2011 Hierarchically Porous TiO2 Electrodes Fabricated by Dual Templating Methods for Dye-Sensitized Solar Cells Adv. Mater. 23 26 2971 2975
5. H. Choi, W. T. Chen, P. V. Kamat, 2012 Know Thy Nano Neighbor. Plasmonic versus Electron Charging Effects of Metal Nanoparticles in Dye-Sensitized Solar Cells ACS Nano 6 5 4418 4427
6. S. Colodrero, A. Mihi, L. Haggman, M. Ocana, G. Boschloo, A. Hagfeldt, H. Miguez, 2009 Porous One-Dimensional Photonic Crystals Improve the Power-Conversion Efficiency of Dye-Sensitized Solar Cells Adv. Mater. 21 7 764 770
7. J. Desilvestro, M. Grätzel, L. Kavan, J. E. Moser, J. Augustynski, 1988 Highly Efficient Sensitization of Titanium Dioxide J. Am. Chem. Soc. 107 10 2988 2990
8. I. K. Ding, J. Zhu, W. Cai, S. J. Moon, N. Cai, P. Wang, S. M. Zakeeruddin, M. Grätzel, M. L. Brongersma, Y. Cui, M. D. Mc Gehee, 2011 Plasmonic Dye-Sensitized Solar Cells Adv. Energy Mater. 1 1 52 57
9. K. Fan, W. Zhang, T. Peng, J. Chen, F. Yang, 2011 Application of TiO2 Fusiform Nanorods for Dye-Sensitized Solar Cells with Significantly Improved Efficiency J. Phys. Chem. C 115 34 17213 17219
10. J. Ferber, J. Luther, 1998 Computer Simulations of Light Scattering and Absorption in Dye-Sensitized Solar Cells Sol. Energy Mater. Sol. Cells 54 1-4 265 275

11. F. E. Gálvez, E. Kemppainen, H. Míguez, J. Halme, 2012 Effect of Diffuse Light Scattering Designs on the Efficiency of Dye Solar Cells: An Integral Optical and Electrical Description J. Phys. Chem. C 116 21 11426 11433
12. X. D. Gao, X. M. Li, X. Y. Gan, Y. Q. Wu, R. K. Zheng, C. L. Wang, Z. Y. Gu, P. He, 2012 Aerogel Based SiO2-TiO2 Hybrid Photoanodes for Enhanced Light Harvesting in Dye-Sensitized Solar Cells J. Mater. Chem. 22 36 18930 18938
13. F. Gao, Y. Wang, D. Shi, J. Zhang, M. K. Wang, X. Y. Jing, R. Humphry-Baker, P. Wang, S. M. Zakeeruddin, M. Grätzel, 2008 Enhance the Optical Absorptivity of Nanocrystalline TiO2 Film with High Molar Extinction Coefficient Ruthenium Sensitizers for High Performance Dye-Sensitized Solar Cells J. Am. Chem. Soc. 130 32 10720 10728
14. M. Gratzel, 2009 Recent Advances in Sensitized Mesoscopic Solar Cells Acc. Chem. Res. 42 11 1788 1798
15. A. Hagfeldt, G. Boschloo, L. Sun, L. Kloo, H. Pettersson, 2010 Dye-Sensitized Solar Cells Chem. Rev. 110 11 6595 6663
16. T. W. Hamann, A. B. F. Martinson, J. W. Elam, M. J. Pellin, J. T. Hupp, 2008a Aerogel Templated ZnO Dye-Sensitized Solar Cells Adv. Mater. 20 8 1560 1564
17. T. W. Hamann, A. B. F. Martinson, J. W. Elam, M. J. Pellin, J. T. Hupp, 2008b Atomic Layer Deposition of TiO2 on Aerogel Templates: New Photoanodes for Dye-Sensitized Solar Cells J. Phys. Chem. C 112 27 10303 10307
18. S. H. Han, S. Lee, H. Shin, H. S. Jung, 2011 Quasi-Inverse Opal Layer Based on Highly Crystalline TiO2 Nanoparticles: A New Light-Scattering Layer in Dye-Sensitized Solar Cells Adv. Energy Mater. 1 4 546 550
19. S. Hore, C. Vetter, R. Kern, H. Smit, A. Hinsch, 2006 Influence of Scattering Layers on Efficiency of Dye-Sensitized Solar Cells Sol. Energy Mater. Sol. Cells 90 9 1176 1188
20. F. Huang, D. Chen, X. L. Zhang, R. A. Caruso, Y. B. Cheng, 2010 Dual-Function Scattering Layer of Submicrometer-Sized Mesoporous TiO2 Beads for High-Efficiency Dye-Sensitized Solar Cells Adv. Funct. Mater. 20 8 1301 1305

21. C. H. Huang, Y. T. Yang, R. A. Doong, 2011 Microwave-Assisted Hydrothermal Synthesis of Mesoporous Anatase TiO2 via Sol-Gel Process for Dye-Sensitized Solar Cells Microporous Meseporous Mater. 142 2-3 473 480
22. N. Hüsing, U. Schubert, 1998 Aerogels-Airy Materials: Chemistry, Structure, and Properties Angew. Chem. Int. Ed. 37 1-2 22 45
23. N. C. Jeong, C. Prasittichai, J. T. Hupp, 2011 Photocurrent Enhancement by Surface Plasmon Resonance of Silver Nanoparticles in Highly Porous Dye-Sensitized Solar Cells Langmuir 27 23 14609 14614
24. J. Y. Kim, S. H. Kang, H. S. Kim, Y. E. Sung, 2010 Preparation of Highly Ordered Mesoporous Al2O3/TiO2 and Its Application in Dye-Sensitized Solar Cells Langmuir 26 4 2864 2870
25. S. H. Ko, D. Lee, H. W. Kang, K. H. Nam, J. Y. Yeo, S. J. Hong, C. P. Grigoropoulos, H. J. Sung, 2011 Nanoforest of Hydrothermally Grown Hierarchical ZnO Nanowires for a High Efficiency Dye-Sensitized Solar Cell Nano Lett. 11 2 666 671
26. H. J. Koo, Y. J. Kim, Y. H. Lee, W. I. Lee, K. Kim, N. G. Park, 2008 Nano-embossed Hollow Spherical TiO2 as Bifunctional Material for High-Efficiency Dye-Sensitized Solar Cells Adv. Mater. 20 1 195 199
27. M. Law, L. E. Greene, J. C. Johnson, R. Saykally, P. Yang, 2005 Nanowire dye-sensitized solar cells Nature Materials 4 6 455 459
28. M. K. Nazeeruddin, A. Kay, I. Rodicio, R. Humphry-Baker, E. Mueller, P. Liska, N. Vlachopoulos, M. Grätzel, 1993 Conversion of Light to Electricity by Cis-X2bis(2,2′-bipyridyl-4,4′-dicarboxylate)ruthenium(II) Charge-Transfer Sensitizers (X = Cl-, Br-, I-, CN-, and SCN-) on Nanocrystalline Titanium Dioxide Electrodes J. Am. Chem. Soc. 115 14 6382 6390
29. M. K. Nazeeruddin, P. Pechy, T. Renouard, S. M. Zakeeruddin, R. Humphry-Baker, P. Comte, P. Liska, L. Cevey, E. Costa, V. Shklover, L. Spiccia, G. B. Deacon, C. A. Bignozzi, M. Grätzel, 2001 Engineering of Efficient Panchromatic Sensitizers for Nanocrystalline TiO2-Based Solar Cells J. Am. Chem. Soc. 123 8 1613 1624
30. B. O. Regan, J. Moser, M. Anderson, M. Grätzel, 1990 Vectorial Electron Injection into Transparent Semiconductor Membranes

and Electric Field Effects on the Dynamics of Light-Induced Charge Separation J. Phys. Chem. 94 24 8720 8726

31. J. J. Pietron, A. M. Stux, R. S. Compton, D. R. Rolison, 2007 Dye-Sensitized Titania Aerogels as Photovoltaic Electrodes for Electrochemical Solar Cells Sol. Energy Mater. Sol. Cells 91 12 1066 1074
32. J. Qi, X. Dang, P. T. Hammond, A. M. Belcher, 2011 Highly Efficient Plasmon-Enhanced Dye-Sensitized Solar Cells through Metal@Oxide Core-Shell Nanostructure ACS Nano 5 9 7108 7116
33. J. Qiu, F. Zhuge, X. Li, X. Gao, X. Gan, L. Li, B. Weng, Z. Shi, Y. H. Hwang, 2012 Coaxial multi-shelled TiO2 nanotube arrays for dye sensitized solar cells J. Mater. Chem. 22 8 3549 3554
34. B. O. Regan, J. Moser, M. Anderson, M. Grätzel, 1990 Vectorial Electron Injection into Transparent Semiconductor Membranes and Electric Field Effects on the Dynamics of Light-Induced Charge Separation J. Phys. Chem. 94 24 8720 8726
35. B. O. Regan, M. Grätzel, 1991 A Low-Cost, High-Efficiency Solar Cell Based on Dye-Sensitized Colloidal TiO2 Films Nature 353 6346 737 740
36. F. Sauvage, D. Chen, P. Comte, F. Huang, L. P. Heiniger, Y. B. Cheng, R. A. Caruso, M. Gratzel, 2010a Dye-Sensitized Solar Cells Employing a Single Film of Mesoporous TiO2 Beads Achieve Power Conversion Efficiencies Over 10% ACS Nano 4 8 4420 4425
37. F. Sauvage, F. D. Fonzo, A. L. Bassi, C. S. Casari, V. Russo, G. Divitini, C. Ducati, C. E. Bottani, P. Comte, M. Graetzel, 2010b Hierarchical TiO2 Photoanode for Dye-Sensitized Solar Cells Nano Lett. 10 7 2562 2567
38. G. B. Shan, H. Assaaoudi, G. P. Demopoulos, 2011 Enhanced Performance of Dye-Sensitized Solar Cells by Utilizationofan External, Bifunctional Layer Consisting of Uniform β-NaYF4:Er3+/Yb3+ Nanoplatelets ACS Appl. Mater. Interfaces 3 9 3239 3243
39. F Shao, J Sun, L Gao, S Yang, J Luo, 2011 Template-Free Synthesis of Hierarchical TiO2 Structures and Their Application in Dye-Sensitized Solar Cells ACS Appl. Mater. Interfaces 6 3 2148 2153

40. N. Tetreault, M. Grätzel, 2012 Novel Nanostructures for Next Generation Dye-Sensitized Solar Cells Energy Environ. Sci. 5 9 8506 8516

41. Y Wei, C Xu, S Xu, C Li, W Wu, Z. L. Wang, 2010 Planar Waveguide–Nanowire Integrated Three-Dimensional Dye-Sensitized Solar Cells Nano Lett. 6 10 2092 2096

42. F. Xu, M. Dai, Y. Lu, L. Sun, 2010 Hierarchical ZnO Nanowire-Nanosheet Architectures for High Power Conversion Efficiency in Dye-Sensitized Solar Cells J. Phys. Chem. C 114 6 2776 2782

43. A. Yella, H. W. Lee, H. N. Tsao, C. Yi, A. K. Chandiran, M. K. Nazeeruddin, E. W. Diau, G. Diau, C. Y. Yeh, S. M. Zakeeruddin, M. Grätzel, 2011 Porphyrin-Sensitized Solar Cells with Cobalt (II/III)-Based Redox Electrolyte Exceed 12 Percent Efficiency Science 334 6056 629 634

44. L. Yang, W. W. F. Leung, 2011 Application of a Bilayer TiO2 Nanofiber Photoanode for Optimization of Dye-Sensitized Solar Cells Adv. Mater. 23 39 4559 4562

45. C. T. Yip, H. Huang, L. Zhou, K. Xie, Y. Wang, T. Feng, J. Li, W. Y. Tam, 2012 Direct and Seamless Coupling of TiO2 Nanotube Photonic Crystal to Dye-Sensitized Solar Cell: A Single-Step Approach Adv. Mater. 23 47 5624 5628

46. J. H. Yum, E. Baranoff, S. Wenger, M. K. Nazeeruddin, M. Grätzel, 2011 Panchromatic Engineering for Dye-Sensitized Solar Cells Energy Environ. Sci. 4 3 842 857

47. T. K. Yun, S. S. Park, D. Kim, Y. K. Hwang, S. Huh, J. Y. Bae, Y. S. Won, 2011 Pore-Size Effect on Photovoltaic Performance of Dye-Sensitized Solar Cells Composed of Mesoporous Anatase-Titania J. Power Sources 196 7 3678 3682

48. Q. Zhang, G. Z. Cao, 2011 Nanostructured photoelectrodes for dye-sensitized solar cells Nano Today 6 1 91 109

49. F. Zhuge, J. Qiu, X. Li, X. Gao, X. Gan, W. Yu, 2011 Toward Hierarchical TiO2 nanotube Arrays for Efficient Dye-Sensitized Solar Cells Adv. Mater. 23 11 1330 1334

Chapter 4

Silicon Solar Cells with Nanoporous Silicon Layer

Tayyar Dzhafarov[1]

[1]Department of Solar and Hydrogen Cells, Institute of Physics, Azerbaijan National Academy of Sciences, Azerbaijan

INTRODUCTION

Today's photovoltaic solar panels are widely used to supply the power and buildings. The account of total solar cell product in 2010 was about 20 GW. Over 95% of all solar cells produced worldwide are composed of the silicon (single crystal, polycrystalline, amorphous, ribbon etc.) and domination of silicon-based solar cell market probably will be do so in the immediate future. The main reason for dominant role of silicon solar cells in word market is high quality silicon that produced in large quantities for microelectronic industry. Additionally silicon solar cell processing does not burden the environment.

The main requirements for ideal solar cell material are (a) direct band structure, (b) band gap between 1.1 and 1.7 eV, (c) consisting of readily available and non-toxic materials, (d) good photovoltaic conversion efficiency, (e) long-term stability [1]. Silicon is the second most abundant element in the earth's crust (35 %) after oxygen. It is the base material for photovoltaic conversion of solar spectrum radiation ranging from ultraviolet to the near infrared, however it can absorb the small portion of solar radiation, i.e. can convert photons with energy of the silicon band gap. The theoretical curve for conversion efficiency of solar cell materials versus band gap for single junction cell (Figure 1) shows that silicon (1.1 eV) is not at the maximum of the curve (about 1.4-1.5 eV) but relatively close to it [2]. The efficiency for ideal silicon solar cell can reach about 30% (for AM1.5 at 300K).

Photoelectron properties related with indirect band structure and high reflectance of crystalline silicon (about of 30-35%) are still a challenger for creation solar cells with high conversion efficiency. High refractive index of crystalline silicon (about 3.5) in solar spectrum region of 300-1100 nm creates large optical losses which can be reduced by using antireflection coating (ARC). Although highly efficient double and triple antireflection coatings are available, most manufactured crystalline silicon solar cells employ simple and inexpensive single-layer ARC with relatively poor antireflection properties.

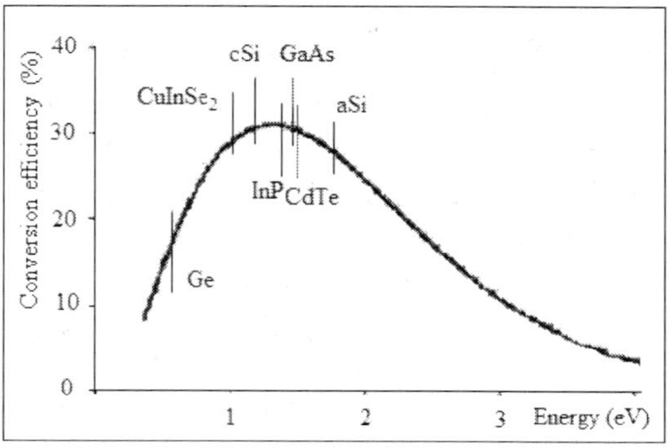

Figure 1: Conversion efficiency depending on semiconductor band gap (AM1.5, 300K).

The first observation of a visible photoluminescence at room temperature in nanostructured porous silicon opened the possibilities of wide range photonic and biologic applications due to tunable refractive index, large surface/volume ratio an biocompatibility of porous silicon [3]. Today the porous silicon is quickly becoming very import and versatile material for solar cell technology.

The crystalline structure, chemical, electrical, photoluminescence and optical properties of porous silicon have been extensively studied by various experimental techniques [4]. Porous silicon can be formed by chemical etching, electrochemical etching and photo-electrochemical etching of silicon in HF-based solutions at room temperature. Therefore the chemical technology can be more adapted to industrial fabrication of solar cells due to its simplicity and lower cost. Porosity, thickness, refractive index of layer, pore size etc. depend on formation parameters (electrolytes contents, current density, temperature, crystal orientation, doping type and concentration, time etching etc.). Sizes of pore and pores walls can be varied from 5-10 nm to hundreds micrometers dependent on fabrication parameters. Possibilities of minimization of reflectance (due to light trapping in pores), increase of band gap of porous silicon layer (due to quantum confinement of charges in the PS micro crystallites) by changing of porosity allow to use PS layer as both ARC and wide-band gap photosensitive layer. Last years the porous silicon layers are widely used in silicon solar cell applications.

This chapter has focused on review of investigations concerning using of porous silicon layers in silicon solar cells and also characterization of structure and properties of porous layers.

PHOTOVOLTAIC CHARACTERISTICS OF SOLAR CELLS

For the solar cell with the series resistance R_s and shunt (or parallel) resistance R_{sh} current-voltage characteristic is determined as [5]

$$I = I_0\left\{\exp\frac{q^{v} - IR_s}{AkT} - 1\right\} + \frac{v - IR_s}{R_{sh}} - I_1 \qquad (1)$$

Here I_o is the reverse saturation current, A is diode ideality factor, q is elementary charge, k is Boltzmann's constant, T is absolute temperature, I_l is photo-generated current. Figure 2 shows the representation of the dark and the illuminated characteristics of the p-n junction.

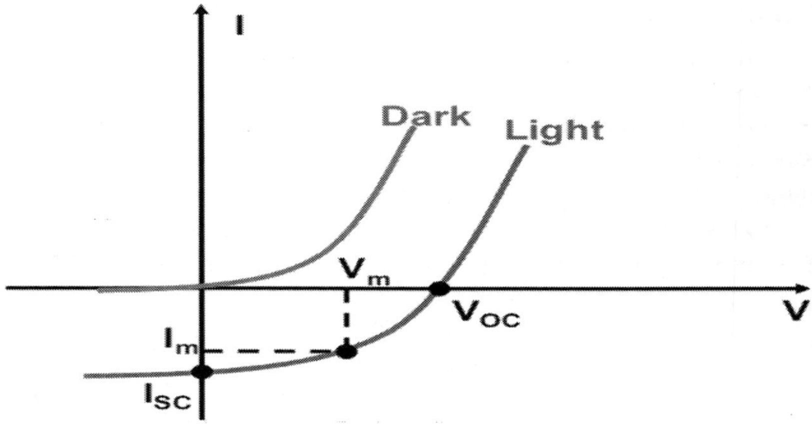

Figure 2: The dark and light current-voltage characteristics of the p-n junction.

The reverse saturation current is given

$$I_0 = \frac{q^{\frac{D_n n_1^2}{}}}{L_n N_A} + \frac{q^{\frac{D_p n_1^2}{}}}{L_p N_D} \tag{2}$$

Here L_n and L_p are diffusion length, D_n and D_p are diffusion coefficient of minority carriers (electrons and holes, respectively), ni is intrinsic carrier concentration, N_A and N_d are concentration of acceptor and donor impurities, respectively.

When the solar cell is operated at open circuit (I=0, i.e. the shunt resistance is high) then the open-circuit voltage is give

$$V_{oc} = \frac{AkT}{q} \ln\left(\frac{I_1}{I_0}\right) \approx \frac{AkT}{q} \ln\frac{I_1}{I_0} \tag{3}$$

Very low values of R_{sh} produces a significant reduction of V_{oc}. An increase in reverse saturation current I_o produces a reduction in Voc. The reverse saturation current is determined by the leakage current of carriers across the p-n junction under reverse bias. The leakage current is a result of recombination of carriers on either side of the junction. For the solar cell working at short circuit (V=0, i.e. low R_s and Io, and high R_{sh}) the short-circuit current is equal to photocurrent $I_{sc} \approx$ Il.

The series resistance of cell depends on concentration of carriers in n- and p-region, depth of p-n junction, resistance and construction of frontal and back ohmic contacts. Increase of series resistance produces a significant reduction in I_{sc}.

The conversion efficiency of the solar cell is defined as the percentage of incident of solar power, which the can convert in electrical power

$$n = \frac{P_m}{E_1 S} \qquad (4)$$

Here $P_m = I_m V_m$ (in Watt) is maximum electrical power, I_m and V_m values of current and voltage corresponding the maximum output power (Fig. 2), El (in W/m²) is light irradiance and S is the surface area of the solar cell.

The fill factor (FF) defines the portion of electrical power produced in solar cell in load. The fill factor is the ratio of the maximum electrical power divided by the open-circuit voltage and the short-circuit current

$$FF = \frac{P_m}{I_{sc} V_{oc}} = \frac{I_m V_m}{I_{sc} V_{oc}} \qquad (5)$$

Substituting P_m from Eq. (4) in Eq. (5) gives for the efficiency

$$n \frac{V_{oc} I_{sc}}{E_1 S} FF \qquad (6)$$

The series and shunt resistance of solar cell influence on the fill factor. Increase of shunt resistance and decrease of series resistance result in to higher fill factor and thereby to larger efficiency.

For crystalline silicon solar cells efficiency about 25% in laboratory and 14% commercially is reached. A theoretical limit of efficiency of crystalline solar cell is about 30%. Comparison efficiency of industrially produced silicon solar cells with theoretical efficiency shows that about 85% power losses occur in commercially cells.

The present efficiency and cost of the silicon solar cell in comparison with conventional energy sources limit the wider using of silicon cells. To improve the performance of solar cells, the power losses must be reduced. The maximum absorption (i.e. minimum reflectance), minimum recombination and series resistance are most conditions for reaching a high efficiency solar cell. The reduction of different energy losses in crystalline silicon solar cells is the most problem of improvement of the conversion efficiency and thereby of reduction of cost.

LOSSES IN SOLAR CELLS

The losses in silicon solar cells can be related with: (a) recombination losses, (b) series resistance losses, (c) thermal losses, (d) metal/ semiconductor contact losses, (e) reflection losses [6].

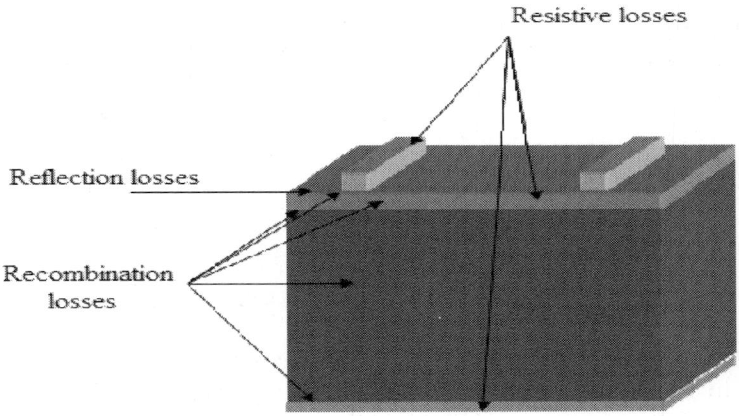

Figure 3: Schematic representation of energy losses in solar cell.

Recombination losses can be caused as result of surface and bulk recombination, recombination in depletion region and recombination at metal/semiconductor contact (Figure 3). Recombination losses mainly influence on the open-circuit voltage.

The incomplete chemical bonds presenting on the surface of semiconductor play role of traps for photo-excited carriers and therefore recombination on traps can result in reduction on photocurrent. The surface recombination velocity S is expressed as

$$S = \sigma \, v \, N_t \qquad (7)$$

Where σ and v are capture cross section for carriers and thermal velocity of carriers, respectively, N_t is the number of surface traps. The decrease of surface recombination velocity is usually reached by deposition of thin passivation films on top surface of cell (for example S_iO_2 or $S_{i3}N_4$ films) by chemical vapor deposition (CVD), plasma enhance chemical vapor deposition (PECVD) or thermal oxidation technique). The standard technique for the reduction of the surface state density of Si is thermal oxidation at 800oC for 15 min in dry oxygen [7]. Passivation of silicon surface results in significant reduction of surface recombination velocity (from 8×10^4 to 1.6×10^2 cm/s).

Impurities and crystalline defects, presenting in bulk region of semiconductor can play role of traps for carriers. Reduction of concentration of rest impurities in bulk of semiconductor, according to Schockey and Read model, will decrease the trap-assisted recombination velocity. Using the bulk semiconductor material having lower concentration impurities and defects can increase the diffusion length of minority carriers and thereby can decrease the recombination losses in bulk of solar cells.

Series resistance of a solar cell consists of several components, such as top grid and bus-bar resistance, emitter resistance; semiconductor bulk resistance and metal/semiconductor contact resistance (see Figure 3). The series resistance controlled by the top contact design and emitter resistance must be carefully designed for each type of solar cell in order to optimize efficiency of solar cell. The semiconductor bulk resistance is low (about 10^{-3} Ω) due to its high conductivity. High values of series resistance will produce a significant reduction in short-circuit current. Losses caused by series resistance are given by $P= V_{Rs} I = I^2 R_s$ and increase quadratic with photocurrent and therefore they most important at high illumination intensities. Very low values of shunt resistance R_{sh} will produce a significant reduction in open-circuit voltage. A low shunt resistance is a processing defect rather than a design parameter. Both series and shunt resistance losses decrease the fill factor and efficiency of a solar cell.

The characteristic equation for solar cell described by Eq. (3) shows that an increase in reverse saturation current I_o produces a reduction in open-circuit voltage. Increase of reverse saturation current means rising of leakage of carriers across the p-n junction under reverse bias due to recombination of carriers in depletion regions on either side of junctions. Recombination proceeding in the depletion regions is less significant as compared to the surface recombination due to the electrical field of p-n junction that separates the photo-generated electrons and holes.

Reduction of emitter layer resistance is reached by optimization of the doping concentration of layer and the p-n junction depth. For (n^+-p) silicon solar cell the optimal values of junction thickness and doping concentration about of $d_{pn} \approx 0.8$ μm and $n_+ \approx 2\text{x}10^{19}$ cm^{-3} respectively.

Thermal losses consist of a significant portion of losses in photovoltaic solar cells. The excess energy formed at absorption of the solar photons that is larger than the band gap energy of semiconductor is realized in the form heat. The temperature rise of solar cell results in increase in the reverse saturation current I_o due to an increase concentration of the intrinsic carriers n_i and diffusion length of minority carriers (Eq. (2)). The increase in reverse saturation current reduces the open-circuit voltage (Eq. (3)). Increase in Io means the rising the "leakage" of carriers across the p-n junction under reverse bias due to recombination of carriers in the neutral regions on neither side of the junction.

Metal/semiconductor contacts placed on the frontal and back surfaces of solar cell cause considerable losses. The screen-printed technique is often used for deposition metallic contacts to silicon solar cells [8]. The frontal contact has form of fine grid lines and the back contact is a metal plate covering entire back surface of cell. Ag and Ag-Al pastes are used in conventional silicon solar cells by deposition of frontal and back ohmic contacts by screen-printed technique. Reduction of resistance of metal/semiconductor contacts is one of major means decrease of power losses for cells. Presence of acceptor type impurity (Al) in back contact (Ag-Al) results in decrease the resistance of near-back contact region of p-type silicon substrate due to diffusion penetration of aluminum during thermal treatment. Moreover, the heavy doping forms the near-surface electric field that reduces the recombination losses at metal/semiconductor contact.

Reflection losses. A large portion of energy losses during solar cell operation could be attributed to optical losses caused by large value of reflectance (about 35%) in the spectral range where silicon is photosensitive [9]. Traditional techniques of texturing and antireflection coating (ARC) have been applied widely to decrease the reflectance of silicon surface. This technology is most used in industrial production of silicon solar cells [10]. The texturisation adds micrometer-scale tilted pyramid structure to the silicon surface, which reduces reflection of the incident light (Figure 4) [11].

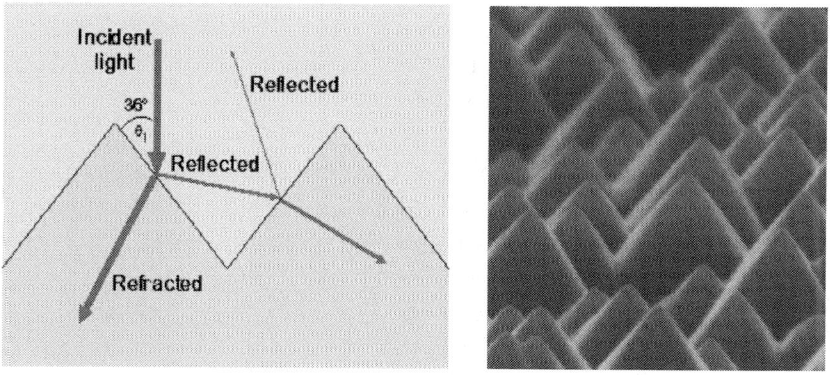

Figure 4: The schematic representation of multiple light reflection (on left) and optical microscopy image of textured silicon surface (on right).

Pyramids are usually formed by etching the surface with acid (H_2SO_4, HNO_3: H_2O etc.) or with alkaline etch (N_aOH, KOH etc.). The light bouncing from pyramid to pyramid increases the optical path and increases absorption of visible light in silicon, thus increasing the efficiency of solar cell. Antireflection coating presents thin film of a transparency material with refractive index (n) between those of air (n_o=1) and silicon (n_{Si} = 3.84). ARC reduces the Fresnel reflection caused from light penetration from a medium of one refractive index to another (for example air/semiconductor).

The optimization of parameters (the refractive index and thickness) of ARC was based on the stratified medium theory and Bridgeman effective medium approximation [12]. The zero-reflection for normal incidence of light on ARC/Si system is given [13]

$$n_{arc} = (n_o n_{Si})\ ½ \tag{8}$$

Here narc, n_o and n_{Si} are the refractive indexes of antireflection coating, the ambient media (air) and silicon, respectively.

The optimal single-layer ARC (SLARC) thickness (d_{arc}) for minimum reflection for wavelength is given [14]

$$d_{arc} = /4n_{arc} \tag{9}$$

If conditions (8) and (9) for air/SLARC/S_i system are to be satisfied (n_o = 1, n_{Si} = 3.84), then the optimal values of refractive index and thickness (a quarter of wavelength) of SLARC must be (for =650 nm) n_{arc} = 1.96 and d_{arc} =80 nm. For glass/SLARC/S_i system (where a glass is protecting layer) optimal values for n_{arc} = 1.55 and d_{arc} = 65 nm.

For double-layer antireflection coating (DLARC) with refractive index of top and bottom layers n_1 and n_2, respectively, on silicon (air/ARC (top)/ARC (bottom)/Si system) the zero-reflectance at normal incident light is realized at conditions [15]

$$n1 = (n_o n_2)½ \text{ and } n_2 = (n_1 n_{Si})\ ½ \tag{10}$$

Here n_1 and n_2 are refractive index of top layer 1 and bottom layer 2, respectively. For air/layer 1/layer 2/S_i system, the ideal values for top layer are n_1 = 1.57 and d_1 = 102 nm, whereas the bottom layer parameters are n_2 = 2.46 and d_2 = 65 nm.

Different types of SLARC (S_iO_2, ZnS, Al_2O_3, and Ta_2O_3), DLARC (MgF_2/ZnS, S_iO_2/T_iO_2, MgF_2/T_iO_2, MgF_2/CeO_2, S_iO_2/S_iH etc.) and multilayer ARCs are used for reduction for reflectance of silicon solar cells. Optimal values refractive index and thickness for single-layer ARCs on silicon surface (air/ARC/S_i system, for = 650 nm) are presented in Table 1.

Table 1: Optimal values of refractive index and thickness of the single-layer ARCs on silicon

ARCs	$_{nar}$c	**d$_{arc}$(nm)**
$S_{iO}2$	1.4	116
$S_{i3N}4$	2	81
Z_nS	2.25	72
Z_nO	2	81
Mg_F2	1.4	116
Ti_O2	2.5	65
Sn_O2	1.9	86
S_iN_x:H	1.9-2.4	68-86
Por.Si	(1.2-2.2)*	74-135

* For porosity from 52% to 80%

As stated above ARCs are generally fabricated by chemical vapor deposition, plasma-enhanced chemical deposition, thermal oxidation processes. They are carried out at high temperatures resulting in an increase in the cost of solar cells [16]. Traditional antireflection coating and surface texturing may reduce reflection efficiently (up to 10-15%) at low wavelength in the visible spectrum (about 400-800 nm), but they are less efficient at harvesting light in the near infrared spectrum (more than 800 nm). The significant of portion of solar radiation, penetrating through the atmosphere, lies at wavelength greater than 800 nm, therefore solar cells with traditional ARC and surface textures leave a huge amount of potential energy out of the using.

Use the single layer ARCs is most simple, low cost and suitable for silicon solar cell technology as compared to expensive and impractical double- or multilayer ARCs. A single layer ARC allows a reduction of reflectance (up to 11%) only in a narrow wavelength range of solar spectrum. A single-layer coating cannot reduce the reflection in a wide wavelength because of neighboring interference maxima. A wider spectral range may be obtained either by increasing the number of layers or by using an inhomogeneous layer with gradient of refractive index. Usage of such inhomogeneous layer allows to suppress the interference maxima narrowing the spectral range. Using the ARC with monotonous changeof the refractive index on the depth can raise the performance of silicon solar cells. ARCs with a graded refractive index constituted from silicon and titanium oxides mixtures were studied in [17]. 3.7% average reflectance between wavelength from 300 to 1100 nm and 48% improvement of the photocurrent was reached on using silicon and titanium oxides mixtures as graded ARC on silicon. It is believed that the porous silicon with tunable refractive index can be adapted production of silicon solar cells due to the simple and cheap technology.

FABRICATION AND PROPERTIES OF POROUS SILICON

Porous silicon layer on monocrystalline Si substrate and its manufacture by the technique of electrochemical etching of silicon substrate in HF solution or by chemical etching in HF- HNO_3 mixture are known as early as from 1956 [3,18]. Electrochemical etching of silicon is attractive because of the possibility to tune the pore size from a few nanometers to a few tens of micrometers, just by choosing wafer doping level and etching conditions. Moreover, a wide range of porous layer thickness, porosities, surface areas and morphologies can be formed depending on the etching conditions. The bulk silicon was shown modifies during the etching to sponge-like structure with silicon columns and hydrogen covered pores.

The simplest electrochemical cell is shown in Figure 5. The Si wafer acts as the anode and the platinum is the cathode. The thickness of porous silicon layer on S_i substrate is determined by duration of etching. The porosity, i.e. the void fraction in the porous layer is determined by the current density (about 10 - 100 mA/cm_2), composition electrolyte, resistance and the doping density of Si substrate.

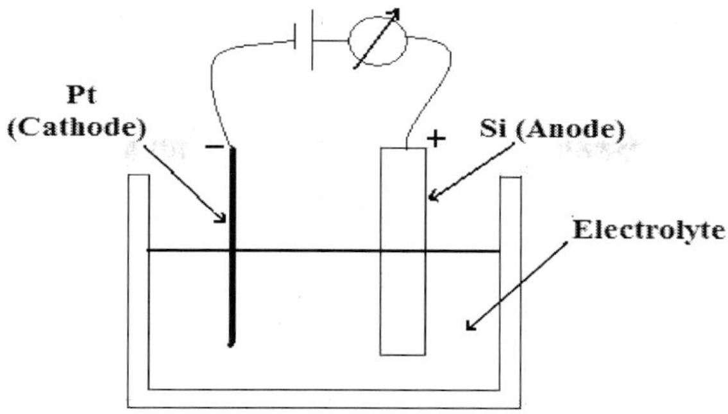

Figure 5: Cross-sectional view of lateral anodization cell.

The anodic reaction on the S_i substrate can be written during pore formation as [19]

$$S_i + 6HF \rightarrow H_2S_iF_6 + H_2 + 2H^+ + 2e^- \quad (11)$$

Silicon atoms are dissolved as S_iF_6 2- require the presence of F- ions (from HF solution) and positively charges holes (from the silicon wafer) at the silicon interface. Concentration of holes in p-S_i is sufficiently high (about 10^{14} - 10^{18} cm^{-3}) and this case the nano-size pores were formed. Concentration of holes in n-Si is very small (about 10^2-10^6 cm^{-3}) and therefore generation of holes is possible due to illumination of n-Si substrate.

The porous silicon layers are often prepared in composition of HF: H_2O, HF: C_2H_5OH, HF: C_2H_5OH: H_2O, HF: HNO_3, HF: HNO_3: H_2O. Fabrication of porous silicon layers on n-type silicon substrates is usually produced under illumination. Before the etching process the silicon substrate are dipped in HF: H_2O (1:50) solution for remove the native oxide film on silicon surface.

The structure and size of pores in porous silicon layer formed on n-S_i substrate differ from those for layer on p-S_i. If electrochemical etching was carried out at relatively low current density (10-80 mA/cm^2), then the local dissolution of silicon surface takes place. Herewith, pore formation begins on surface defects of S_i and further growth of pores into silicon substrate proceeds due to the holes diffusion to Si-electrolyte interface. In the case of large current density (0.5 - 0.8 A/cm^2) when the amount of holes moving to Si-electrolyte interface is very high, the etching of top regions of Si substrate is preferred. It ensures the uniform etching of silicon surface and formation a smooth surface of substrate (the so-called the electro-polishing process). The raising the current density above the critical value at the end of anodization process results in a detachment of the porous silicon film from S^i substrates. The behavior at high current densities turns out to be useful to produce porous silicon free-standing layers. The anodic reaction during the electro-polishing can be written as

$$Si + 6HF \rightarrow H_2S_iF_6 + 4H^+ + 4e^- \quad (12)$$

Pores, depending on the diameter, denoted as micropores (R < 2 nm), mesopores (2 nm < R < 50 nm) and macropores (R > 50 nm). Under illumination the pore size dependent on doping density and anodization conditions, with diameters in the range 100 nm - 20 μm (macropores).

The average porosity (P), i.e. the avoid fraction in the porous layer, can be obtained by gravimetric using the equation

$$P = \left\{ \frac{m_1 - m_2}{m_1 - m_3} \right\} 100(\%) \quad (13)$$

Here m_1 is S_i sample mass before the anodization etching, m_2 just after etching and m_3 after the removal of the porous layer by electro-polishing or after a rapid dissolution of the whole porous layer in a 3% KOH solution. Guessing the porous silicon mass m_{PS}, the average porosity can be also determined by using the equation

$$P = \frac{1 - m_{PS}}{\rho^{sd}} = \frac{m_1 - m_{ps}}{m_1} \quad (14)$$

One can also get the porous silicon layer thickness d using the equation

$$d = \frac{m_1 - m_3}{\rho^s} \quad (15)$$

Here is the S_i density (2.33 g/cm^3) and S_i s the etched surface.

The inhomogeneity in porosity and thickness of porous of the porous layers is often observed on fabrication with electrochemical anodization cell. They are most probably due to the bubbles that form and stick on silicon surface. The inhomogeneity in porous and thickness must be removed and the concentration of the HF has to be locally constant on the surface of the silicon substrate. Removal of the bubbles on the surface of the silicon and thereby preparation of homogeneous porous silicon layers is realized with using a stirrer. The distance between the silicon wafer and the platinum cathode also influences on the homogeneity, whereas the shape of platinum cathode almost does not influence on homogeneity. There is a certain distance for given cell when the porous silicon layers are homogeneous.

The thickness of the porous silicon layers mainly depends on duration of anodization process, whereas the porosity depends on the current density. It is be noted that the character of the thickness-etching time and porosity-current density relations depend on orientation, type and concentration level in silicon and conditions anodization process (the electrolyte composition, distance between silicon wafer and platinum electrode, illumination etc.).

The thickness-etching time dependence for porous silicon layer fabricated on p-type (100) silicon substrates in $3HF:1C_2H_5OH$ solution at current density 20 mA/cm^2 is given in Figure 6 [20]. It is seen that the average growth kinetics of porous silicon layer is about 14 nm/s.

Figure 6: Thickness vs. etching time for porous silicon growth at constant current density of 20 mA/cm^2.

The porosity of the porous silicon layer almost linearly increases with the current density once the other etching parameters are kept fixed (Figure 7).

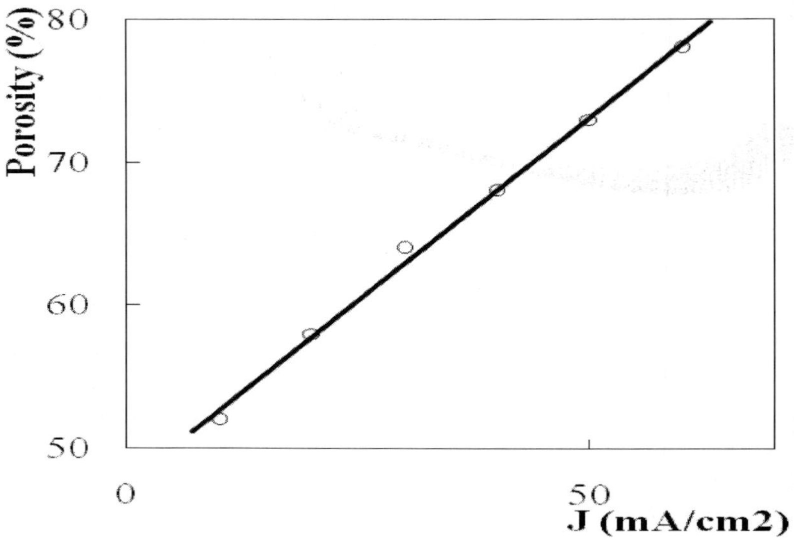

Figure 7: The porosity as a function of current density.

Porous silicon is a particular form of crystalline silicon. The crystalline structure of porous silicon presents a network of silicon in Nano (micro)-sized regions surrounded by void space with a very large surface-to-volume ration (up to 10^3 m^2/cm^3) [13]. The structure of porous silicon is like a sponge or columnar where quantum confinement effects play fundamental role [4]. The pore surfaces are covered by silicon hydrides (Si-H) and silicon oxides (Si-O).

Figure 8 shows Fourier transform infrared (FTIR) spectrum of free-standing PS film of thickness of 12 μm measured at room temperature [21]. The peaks related with absorption on vibration of Si-H (2100 cm^{-1}) and Si-O bonds (1100 cm^{-1}) located on pore surfaces were observed from Figure 8. These bonds play an important role in regulating optical, electrical and gas sensing properties of porous silicon.

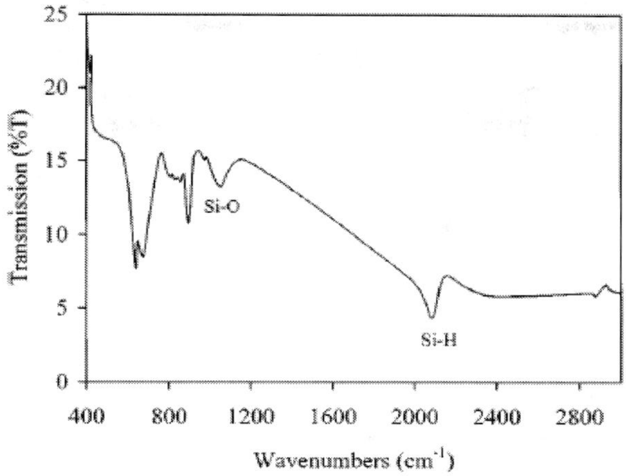

Figure 8: FTIR spectrum of porous silicon film (300 K).

The effect of isothermal annealing of free-standing PS films on changes of intensity of absorption coefficient of Si-H (2100 cm^{-1}) and Si-O (1100 cm^{-1}) peaks is used for estimation of diffusion coefficient [22]. Results of these measurements showed that in the range of 65-185°C the temperature dependence of hydrogen and oxygen diffusion coefficient along the porous surfaces are described as

$$D\,(H) = 5x10^{-10} \exp\,(-0.37\ e^{V}/kT) \qquad (16)$$

$$D\,(O) = 1.3x10^{-8} \exp\,(-0.50\ e^{V}/kT) \qquad (17)$$

The activation energy for diffusion of hydrogen along the porous surfaces estimated from response (or recovery) $V_{oc}{}^{-t}$ curves for Au/PS/Si cells under humid ambient (90%RH) is 0.34 e^{V} [23].

The thermal oxidation of free-standing PS films in the range from 400 to 900°C is accompanied the structural phase transition [24]. The crystalline nanostructured silicon partly converts into amorphous and polycrystalline silicon, if temperature is about 500°C. At higher temperatures three Si structures (crystalline, polycrystalline and amorphous) produce SiO_2 (combination of cristobolite and quartz) due to the oxygen diffusion and absorption in PS. An optimal oxygen-absorption temperature is about 700°C.

The characterization of the lattice deformation of porous silicon carried out by X-ray diffraction has been described in [25, 26]. Crystalline structure of porous silicon layers is equivalent to that of nearly perfect silicon. Porous silicon may be considered as an assembly of small silicon crystallites. These crystallites have two different dimensions, the bigger one being oriented perpendicular to the surface. Typical values seem to be about 1000 Å and 10-100Å. Lattice parameter of the porous silicon of 54% porosity prepared on p^+-Si (100) substrate was slightly bigger (the difference (Δa) is 2.3×10^{-3} Å) than that of intrinsic silicon (a= 5.4306 Å).

A linear increase in the lattice parameter expansion (the lattice mismatch parameter $\Delta a / a$) for porous silicon going from 4 to 7×10^{-4} when the porosity increases from 34 to 72% has been measured [25]. The same behavior for lattice mismatch parameter $(\Delta a / a) \times 10^4$ between porous silicon layer and p^+-type silicon substrate depending on porosity has been found in [27].

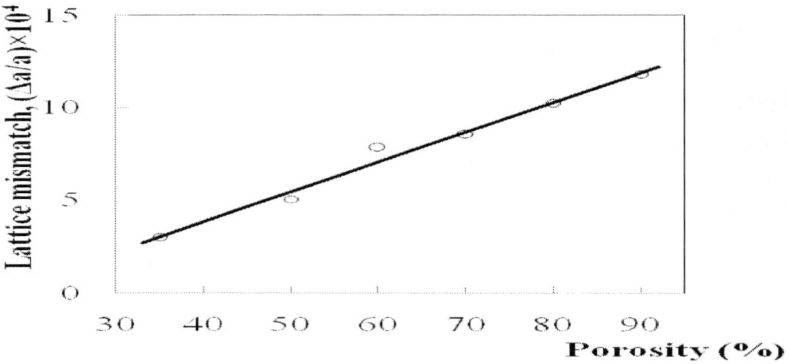

Figure 9: Relation for lattice mismatch parameter $\Delta a/a$ between the PS layer and p^+-type Si substrate.

The origin expansion is attributed to the hydrogen-silicon bonds at the inner surface of the porous silicon. The hydrogen desorption results in a sharp contraction of the lattice parameter of porous silicon layer.

The pores on the surface of silicon increases absorption of light by increasing the effective thickness of emitter layer of solar cell (see Figure 4). Therefore the refractive index of the emitter layer, which strongly influences on efficiency of the solar cell depends on porosity of porous silicon. The current density applied during the formation process determines the porosity of the porous silicon layer and consequently its refractive index. We can say that the porosity-current density profile is transferred to the refractive index versus current density profile.

The reflectance can be calculated from refractive index. For film with parallel surfaces when light moves from a medium with refractive index n_1 to one with refractive index n_2 reflectance for normal incident is given as

$$R = (n_1 - n_2/n_1 + n_2)^2 \quad (18)$$

The reflectance spectrum of the porous silicon film is characterized by the multiple interference fringes caused by the air-porous silicon and porous silicon-silicon interfaces. A simple method for evaluating the refractive index on a thin film is to measure the interference fringes resulting from multiple reflections

$$n = 1/2d\,(_{1}\,_{2}/\,_{2} - \,_{1}) \quad (19)$$

Here $_1$ and $_2$ are the wavelength for two consecutive maxima, d is the thickness of the film. The main advantage of this method determination of the refractive index is its rapidity and simplicity. This method has been used by different authors for normal incidence. Figure 10 illustrates the refractive index as a function of current density for low resistivity silicon substrate [28].

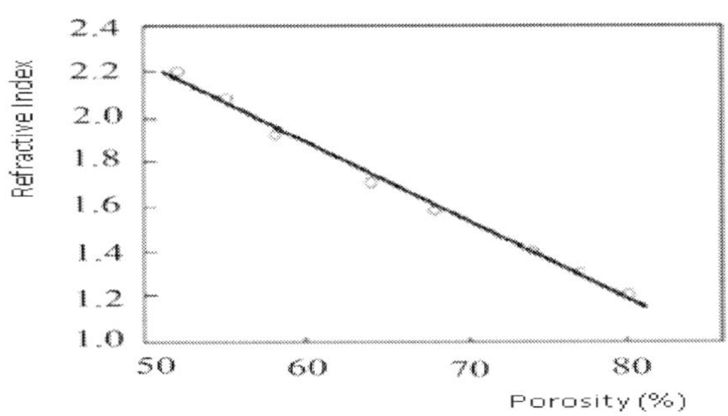

Figure 10: Refractive index as a function of current density for p^+-type doped substrate.

Three type of porous silicon layer with different refractive index profile along the thickness are used in solar cells as antireflection coating: (a) the refractive index is constant, (b) the refractive index profile smoothly changes, and (c) it changes in stepped form. The (b) type of porous silicon layer is fabricated by smooth change the current density during anodization process and (c) type porous silicon presents the multilayer structure which can be prepared by stepped variation of the current density during growth process.

As stated above the porous silicon consists of a network of nano-sized silicon walls and voids that formed when crystalline silicon wafer are etched electrochemically in HF-based electrolyte. Porous silicon presents the quantum system in which the charge carriers located in narrow crystalline silicon wall separating the pore walls. One of features of nanoporous silicon in comparison to the bulk silicon is shifting of fundamental absorption edge into the short wavelength of the visible region of the solar spectrum. It was confirmed by the measuring the optical absorption in the free-standing porous silicon layers [29].

PS layers with a thickness of 10-20 μm and an average porosity of 40 to 80% were prepared on n-type (111) Si substrates ($=1\times10^{-2}$ Ω cm) by anodic etching in HF: H^2O = 1:3 solution at a dc current of about 10-60 mA cm^{-2} under white-light illumination [30]. For optical and electrical measurements, the PS films were then detached from the Si substrate by electro- polishing in the same solution with a current density of 0.8-1.0 a cm^{-2}. The free-standing PS films were characterized by porosity, thickness, optical and resistance measurements. The average porosity was measured by a gravimetric technique. Resistance and charge carrier concentration measurements were carried out on the free-standing PS layers attached to a dielectric substrate (glass) by using the Van der Pauw technique.

Figure 11 shows the effective energy gap in dependency on porosity of the free- standing PS films, calculated from extrapolation of the high energy part of { ^{2} (h) ^{2}- h } spectra [29]. Near linear increase of band gap from 1.4 to 1.9 e^V with rising of porosity of PS films in the range of 30-90 % is observed.

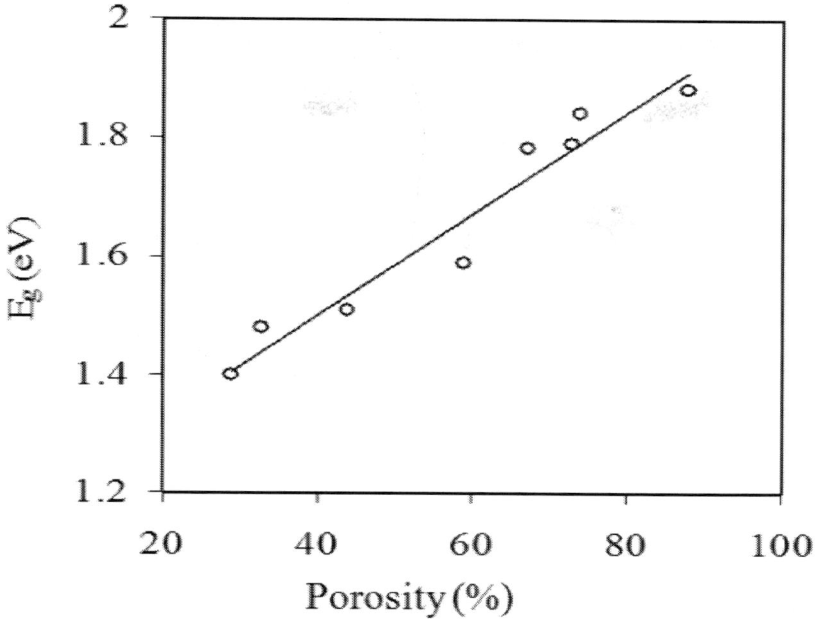

Figure 11: Energy band gap in depending on porosity of PS film (40% RH, 300 K).

Data on Figure 11 concerning increase of the energy gap in dependency on porosity of PS films can be explained by a model including the quantum confinement of carriers in the PS micro crystallites, causing the widening of the S_i band gap.

The electrical measurements of the free-standing PS layers with 65% porosity (300 K, 45% RH) gave values of $= 1.8\times10^6$ Ω cm for resistance, $p = 9.6\times10^{12}$ cm^{-3} for hole concentration, and $\mu = 0.36$ $cm^2/(V\ s)$ for hole mobility.

POROUS SILICON LAYERS IN SILICON SOLAR CELLS

As stated above the features of porous silicon (a quantum system, a sponge or columnar structure and an extremely large pore surfaces) provide many possible applications, such as light emitting diode, chemical and biological sensor, hydrogen fuel cell, photovoltaic cell, antireflection coating in solar cells etc.

Decrease of reflectance (between 30 and 3%) and increase of band gap of porous silicon layer (between 1.1 and 1.9eV) with increase of porosity makes nonporous silicon as a promising material for use in the solar cell technology. Therefore formation of nonporous layer on frontal surface of PS/Si solar cell with lower reflectance and larger band gap, expanding the spectral range of photosensitivity, will contribute to increasing of conversion efficiency. Moreover, formation of Si-H and Si-O bonds on silicon surfaces followed by electrochemical etching in HF-based solution will provide passivation the pore surfaces.

Thus, the room temperature fabrication only nonporous silicon layer on frontal surface of ready silicon solar cell, instead of three-step process (texturization, antireflection layer deposition and passivation), performed at high temperatures on standard technology can essentially improve the photovoltaic parameters and decrease the cost of silicon solar cells.

The potential advantages of porous silicon in silicon solar cells include:

1. Use as antireflection coating due to a lowering of the reflectance in the sensitivity range of silicon solar cell and possibility of formation PS layer with smooth change the refractive index between those Si and air
2. Use as a wide-band optical window (the band gap shifts from 1.1 eV to 1.9 eV as the porosity is increased from 30 to 85% [29,31] that can broad the photosensitive region of the solar cell
3. Use as front semiconductor layer with a variable band gap that can result in increase of photocurrent
4. Possibility of the conversion of high energy ultra-violet and blue part of the solar spectrum into long wavelength radiations due to photoluminescence in Nano crystalline porous silicon
5. Surface passivation and guttering role of porous silicon [32]
6. Simplicity and lower cost fabrication technology of nonporous silicon due to electrochemical modification of silicon

The theoretical requirements for the design of single- and double-layer porous silicon as antireflection coating on silicon solar cells are given in [33]. It is shown that the effective reflectance of $(n^+\text{-}p)$Si solar cell with shallow n^+-emitter depth (about 0.3 μm) can be reduced to 7.3% with a single porous silicon layer (of 80-90 nm thickness) formed a few seconds under current density of 50 mA/cm^2. Taking account of the possibility to modulate optical properties of porous silicon by changing the electrochemical parameters during formation (duration of etching, current density etc.) to reduce the reflectance, structural parameters of double-layers porous silicon are calculated and lower reflectance double-layer porous silicon ARC is realized in [33]. The lowest value of the effective reflectance (below 3%) of a double-layer PS ARC on (n^+-p) Si cell with 0.5 μm-thick emitter have been obtained under 1 mA/cm^2 for 1 s (bottom layer of 47 nm thickness and 37% porosity) and under 50 mA/cm2 for 100 s (top layer of 77 nm thickness and 71% porosity). It is experimentally shown that multiple PS layers ARCs can be formed in a single step procedure by changing the current density during its electrochemical preparation.

The significant reducing of the effective reflectance (up to 3%) was observed for monocrystalline silicon with porous silicon layer formed on previously texturized surface of sample [34]. Porous silicon layer was fabricated by electrochemical or chemical etch (stain etching) in HF: HNO_3: H_2O for 3-60 s on p-type monocrystalline silicon (C_z) or multicrystalline silicon samples. The monocrystallne wafers were previously texturized by anisotropic etching in alkaline solution.

Data of integral reflectance (for = 650 nm) of silicon samples without and with texturized layer, porous silicon layer or antireflection layer are presented in Table 2. The main conclusion which can be made from Table 2 is the effective reflectance for silicon samples with porous silicon layer is significantly smaller than that for samples without porous silicon layer. Moreover, the minimal effective reflectance (about of 3 %) is reached for porous silicon layer formed on previously texturized surface of silicon. The efficiency of PS/(n^+-p)Si solar cell with PS formed after phosphorus diffusion (12.1%) is larger compared to reference silicon solar cell without PS layer (9.4%).

Table 2: The effective reflectance for different silicon surfaces [34]

Samples	Effective reflectance (%)
Polished surface	33.9
mc-Si with ARC (SnO_x)	9.4
PS by stain etching (mc-Si)	6.6
Texturisation by alkaline etching (Cz-Si)	12.6
Texturisation by alkaline etching + PS by electrochemical etching (Cz-Si)	3.4
Texturisation by alkaline etching + PS by stain etching (Cz-Si)	3.1
PS by electrochemical etching (Cz-Si polish surface)	9.7

Antireflection properties of nanoporous silicon layer on *p*-type silicon were investigated in [36]. PS layers were prepared by electrochemical etching of silicon in 1HF: $1C_2H_5OH$ solution. The average reflectance between wavelengths 300-1000 nm was 10.3 % for the optimal PS layer. The analysis of the internal quantum efficiency of *(n^+-p)* silicon solar cell with porous silicon layer as antireflection coating showed that quantum efficiency was comparable to that of a solar cell with a SiN_xantireflection coating prepared using plasma-enhanced chemical vapour deposition.

There are two technology of formation of porous silicon layer on silicon solar cells: (1) the thin porous silicon is formed on final step on surface of ready Si solar cell with metal contacts and (2) the relatively thick porous silicon layer is formed prior to emitter diffusion and metal contact deposition. In the first case the thickness of porous layer (70-150 nm) must be less than depth of n^+-*p* (or p^+-*n*)-junction (300-800 nm), $d_{PS} < dpn$ and duration of electrochemical etching is short (about 5-15 s). In the second case the thickness of the porous layer (5-15μm) is significantly large than the depth of location of n^+-*p* (or p^+-*n*) junction ($d_{PS}>d_{pn}$) and duration formation of porous layer is significantly larger (about 10-30 min). It can be expected that the profile of n^+-p junction must be a flat in the first case and non-flat (it is similar to profile of porous layer surface) in the second case.

At first bellow will be considered results on photovoltaic properties of silicon solar cells with thin porous layer formed on final step fabrication of cells.

The surface modification of silicon solar cells was used for improvement of photovoltaic characteristics of silicon solar cells in [37]. *p*-type boron-doped monocrystalline silicon wafers with orientation of (100), resistivity about of 3 Ω cm and thickness of 250-380 μm were used for fabrication of solar cells by screen-printed process [8, 38]. n^+-emitter layer with 0.5-1.0 μm thickness and 15-20 Ω/□ sheet resistance was formed as a result of phosphorus diffusion. Formation of porous silicon layer on n^+-surface of the device was performed on the final step of the solar cell fabrication sequence. Fabrication of PS layer on n^+-p junction was carried out at galvanostatic condition (constant current) in an electrolyte solution $HF:C_2H_5OH:H_2O$ (1:1:1 in volume) under illumination. Choice of optimal thickness of porous silicon layer as ARC on surface of (n^+-p) silicon solar cell and choice of the refractive index, which strongly depends on porosity (see Figure 9), were defined from conditions presented above.

If conditions (8) and (9) for air/ARC/Si system are to be satisfied (n_0=1 for air and n_{Si}=3.84 for Si), then the optimal values of refractive index and thickness (a quarter of wavelength) of the porous silicon layer, acting as ARC must be (for = 650 nm) n_{arc}=1.96 and d_{arc}=83 nm, respectively. Taking into account the refractive index depends on porosity of porous silicon (Figure 9) one may conclude that the porous silicon layer of 80-90 nm thickness and about 55% porosity (n=2) may act as ARC having minimum reflectance, that in turn will improve the photovoltaic parameters of PS/(n^+-p)Si solar cells.

The growth rate of porous silicon on Si substrate, measured in this run for a current density of 60 mA/cm^2, was about 8 nm/s. Therefore, the time of electrochemical etching under a constant current of 40, 50 or 60 mA/cm^2 was 8-15 seconds. As a result, a blue colored PS layer between the grid fingers on the surface of the n^+-emitter silicon solar cell has been obtained. For some measurements, the PS layers were then detached from the Si substrate by electropolishing process [30]. Free-standing PS layers were characterized by porosity, thickness, and resistivity, photoluminescence and reflectance measurements. Resistivity measurements, carried out by the Van der Pauw technique on the free-standing porous silicon layer of 60% porosity, gave $3x10^4$ Ω cm.

A SEM micrograph of the front PS surface obtained by using scanning electron microscopy (SEM) (JSM-5410LV) is shown in Figure 12. Cross cut representation of silicon layer showed that the pores have a conical form.

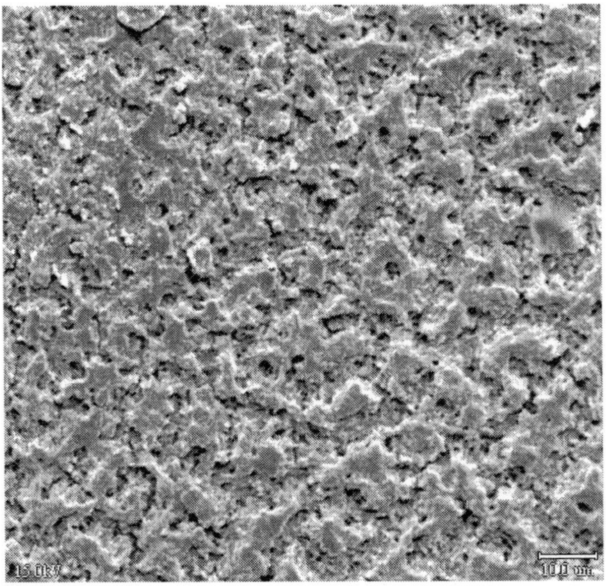

Figure 12: Scanning electron microscopy of porous silicon layer surface.

Figure 13 shows the photoluminescence spectrum of the PS layer (60% porosity) on Si substrate, where the spectrum illustrates the peak at =580 nm (the orange region of the solar spectrum). Measurements of distribution of photoluminescence intensity along the thickness of the PS layer (of thickness 10 μm) showed that the intensity approximately linearly decreases from the surface deep down. Similar results were also obtained by investigations of samples with PS layers of different thicknesses. Observation of photoluminescence in PS at visible region of the spectrum can be interpreted by quantum confinement effect causing the confinement of the charge carriers in Nano crystalline silicon wall separating the pore [39].

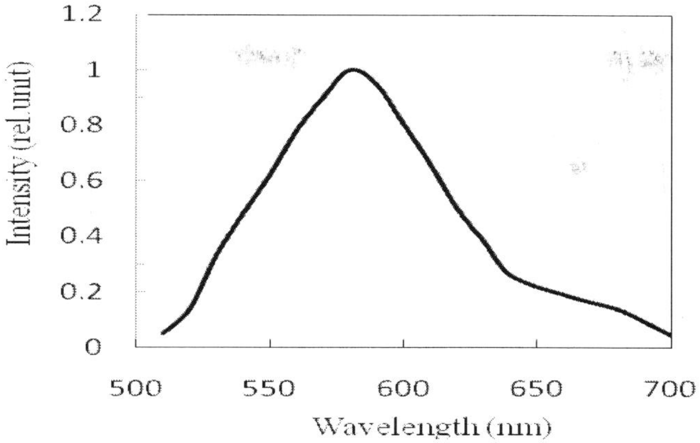

Figure 13: The photoluminescence spectrum for porous silicon layer.

The integrated reflectance spectra of the polished silicon surface before and after porous silicon layer (of 60% porosity) formation showed the significant lowering of the reflectance (to 4%) as compared to polished silicon (about 38-45%). These data show that PS on (n^+-p) silicon solar cell can be used as effective antireflection coating.

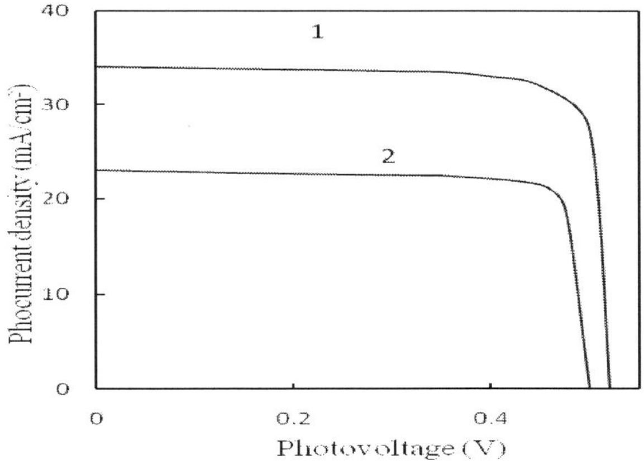

Figure 14: Photocurrent density-voltage characteristics of (n^+-p) silicon solar cell (1) with and (2) without PS layer.

The current-voltage characteristics of n^+-p silicon solar cells without and with porous silicon layer of 60% porosity ((n^+-p) Si and PS/ (n^+-p)

Si structures respectively), measured under AM1.5 illumination, have been presented in Figure 14. As a result, due to the PS coating, the short-circuit current density increased from 23.1 to 34.2 mA/cm^2 and the open-circuit voltage increased from 500 to 520 mV. The experimental results on the photovoltaic parameters for thirty solar cells before and after formation of the PS layer on n$^+$-emitter surface showed that the mean increment of photocurrent density is about 48% (Table 3). At the same time, the open-circuit voltage increase is about 4%. The fill factor remains approximately the same for cells with porous layer (FF=0.75) as compared to the solar cells without porous layer (FF=0.74). The mean efficiency of photovoltaic conversion for solar cells with PS layer increased from 12.1 to 14.5%, what equals a relative increment of about 20%.

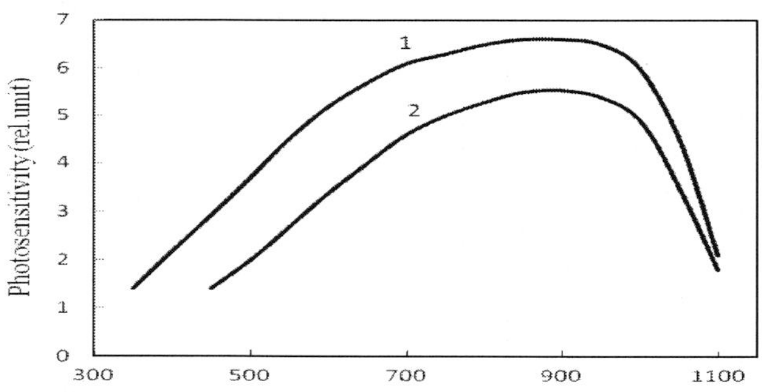

Figure 15: The photosensitivity spectra of (1) PS/ (n^+-p) Si and (2) (n^+-p) Si solar cells.

The photosensitivity spectra of the solar cells with and without PS layer were presented in Figure 15. The value of photosensitivity for the PS/ (n^+-p) Si cell is larger (by about 25%) and the spectral photosensitivity region is considerably wider than those for (n^+-p) Si cell without the PS layer. The significant improvement of photovoltaic characteristics of the PS/ (n^+-p) Si solar cell was attributed mainly to the two-fold role of the porous silicon layer on surface of the n^+-emitter. Firstly, it behaves as a antireflection coating, increasing the incident photon flow on the p^+-n junction and secondly, it plays the role of a wide-band gap optical window (about of 1.8-1.9eV for PS of 60% porosity), broadening the spectral region of photosensitivity of the cell to the ultraviolet part of solar spectrum.

Table 3: Photovoltaic parameters of silicon solar cells without and with porous silicon layer

Solar cell	J_{sc}(mA/cm^2) or I (mA)	V_{oc}(mV)	FF	Eff. (%)	Reflectance (=650 nm)	Reference
(n^+-p) SiPS/(n^+-p) Si	23.134.2	500520	0.740.75	12.114.5	284	[37]
(n^+-p) SiPS/(n $^+$-p) Si	21.528.4	580585	0.550.74	7.512.5	123	[40]
(n^+-p) SiPS/(n $^+$-p) Si	95 mA137 mA	580570		10.313.5		[41]
(n^+-p) SiPS/(n $^+$-p) SiText. (n^+-p) SiPS/Text. (n^+-p) Si	17.220.123.325.5	598606592595	0.740.750.700.74	7.69.59.611.2	7123	[42]
(n^+-p) SiPS/(n $^+$-p) Si	18.527.2	580601	0.730.77	7.8512.54	7	[43]
PS/ (n^+-p) SiSiO$_x$/(n^+-p) Si	33.434.8	460530			93.8	[44]
(p^+-n)PS/Si (100) (p^+-n)PS/ Si (111)	15.9 mA12.4 mA	480440	0.810.82	15.411.2	716	[45]
(n^+-p)PS/Si (111) SiO$_2$/(n^+-p) Si (111) (ZnO-TiO$_2$)/(n^+-p)Si	12.4 mA6.04 mA10.2 mA	440370410	0.820.790.81	11.24.48.4		[47]
(p^+-n)PS/SiPS on one side (p^+-n)PS/Si/PS PS on both sides	8.8 mA 12.4 mA	430 490	0.78 0.84	7.4 12.75	16 6	[48]
(n^+-p)PS/Si	28.9	627	0.76	13.8	9	[49]
PS/(n^+-p) SiSiN/(n^+-p) Si	26.3 28.4	602 606	0.760.75	12 13	10	[50]
(n^+-p) mc-SiPS/ (n^+-p) mc-Si	26.6 28.9	572 582	0.750.76	11.3 12.7	155	[51]
(n^+-p) mc-Si (n^+-p)PS/mc-Si	29.8 30.2	577 587	0.5 0.76	12.913.5	8	[7]

Change the porosity deep down to PS layer can also stimulates the improvementofthephotovoltaicparametersofsolarcells. Experimentally observed decrease of intensity of the photoluminescence peak at 580 nm deep down to PS layer, which consists of pores of conical form, can be circumstantial evidence for decrease of porosity deep down. Taking into account that the band gap energy of nanoporous silicon increases with increment of porosity due to quantum confinement of carrier charges (see Figure 11), one can assume that the porous silicon layer on the (n^+-p) Si cell is a semiconductor with variable band gap width (changing from about 1.8-2.0 eV on the front PS surface to 1.1 eV on the PS/n^+Si interface). As a result, the internal electrical field of the porous layer with variable band gap can also increase of the photocurrent generated in the PS/ (n^+-p) Si solar cell.

The original formation technique of porous silicon layer on silicon solar cells was used in [40]. n^+-$p$junctions were obtained by phosphorus diffusion into monocrystallinep-type silicon wafers. PS layer was formed after deposition of front grid contact. New method PS layer formation consisted in deposition a thin Al film on front surface silicon cell and then HF-spray-etching. Advantage of method is use HF solution only, instead HF:HNO_3 composition, since HF-spray-etching does not influence on front grid metallic contact. Hydrogen-rich PS prepared by HF-spray-etching method plays role of a passivation layer and decreases the surface reflectance from 12% for textured surface to 3% for textured surface in the presence of a thin porous silicon layer. Formation of PS layer prepared under optimized HF-spraying conditions improves the conversion efficiency from 7.5 to 12.5% (Table 3). The observed results were discussed throughout increase of light absorption due to reflectance lowering and band gap widening. Unfortunately, data on thickness, porosity and other parameters of porous layers are not presented in [40].

The porous silicon as ARC in (n^+-p) silicon solar cells was used in [41]. Thin porous silicon layer (of thickness about 100 nm) on n^+-emitter was formed on final stage of (n^+-p) solar cell fabrication. 1HF: 1C_2H_5OH and 1HF: 1C_2H_5OH :H_2O solutions were used as electrolyte during anodization modification of silicon (for 3-12 s). n^+-p junction with n^+-emitter of 400 nm thickness was formed by phosphorus diffusion in monocrystalline p-type (100) silicon of 1.5 Ω cm resistivity. The reflection spectra of porous layer prepared on the polished surface of n^+-emitter showed that the average reflectance is about 7.6% in the wavelength range 400-700 nm with minimum value of 1.4% at 580 nm. The photovoltaic parameters of (n^+-p) Si solar cells before and after formation of PS layer are presented in Table 3. It is seen that the efficiency of solar cell increased from 10.3 to 13.5% as result of fabrication of porous layer on front surface of the cell. The significant increasing for the short-circuit current (from 95 to 137 mA) and very small decreasing of the open-circuit voltage were observed for cells with porous silicon coating. Authors [41] supposed that the formation of porous silicon layer on silicon solar cells does not ensure necessary level of passivation and also does not have sufficient temporal stability.

The photovoltaic parameters of (n^+-p) Si solar cells with and without thin porous silicon layer prepared by standard screen-printed technique are studied in [42]. *P*-type Si (100) wafers of 1.5 Ω cm resistivity and 20 cm^2 surface were used. Formation of porous silicon layer on n^+-surface of cells was performed by electrolytic anodization in 1HF: $4C_2H_5OH$ solution with a current density of 10 mA/cm^2during short time 10 s. Data on characteristics of solar cells, prepared on base of polished monocrystalline and textured monocrystalline silicon without and with porous layer are presented inTable 3. The main conclusion from results presented in Table 3 consists in increasingthe short-circuit current density and efficiency of silicon solar cells with porous layer. For both group cells prepared on polished and textured silicon with porous layer efficiency increases by about 20-25% compared to a cells without porous silicon layer. However the open-circuit voltage and fill factor for cells with porous layer remain almost same with those for cells without porous silicon. The minimum reflectance and the best photovoltaic parameters have the solar cells with porous silicon layer prepared on textured surface.

A significant improvement in efficiency of (n^+-p) Si solar cells with the thin porous silicon layer has been achieved in [43]. Float-zone (100) oriented *p*-type silicon with 1Ω cm resistance was used as starting material. Formation of the porous silicon layer was made on finished solar cell during chemical etching in HF: HNO_3:$H_2$0 solution for 20 s. The reflectance of polished Si sample is significantly lowered after porous layer formation from 36 to 9.5%.The silicon solar cell with porous layer leads to a 46% and 38% relative improvement in photocurrent density and efficiency respectively (Table 3). The observed enhancement in the photovoltaic parameters was attributed to both the antireflection and passivation properties of porous silicon.

The passivation capabilities of PS accompanied with lowering of recombination velocity of charge that can arise from the presence of Si-hydrides and Si-hydroxides on pore surfaces.

Double layers SiO_x/PS structure as antireflection coating in (n^+-p) Si solar cells prepared the screen-printing technique was studied in [44]. Then porous silicon layers were fabricated on the front surface of screen-printing (n^+-p) Si solar cells by electrochemical etching in 1HF:3C_2H_5OH solution at a current density of 20 mA/cm^2 for 4-12 s. The silicon oxide films of different thickness (70-105 nm) were deposited on porous silicon by PECVD method. Decrease of reflectance and improvement of photovoltaic parameters of double layer SiO_x/PS/ (n^+-p) Si cells were received compared to the single PS/ (n^+-p) Si cells. Best results were obtained when the thickness of SiO_x film and PS layer (P=60%) are 105 and 60 nm respectively (Table 3). An improvement in the open-circuit voltage of 14% for double-layer ARC authors [44] attribute to enhancing the surface passivation of the pores due to diffusion of hydrogen from SiO_x film.

There are a number of studies on silicon solar cells with thick porous silicon coating in which the porous layer is formed before fabrication of n^+-p (or p^+-n) junction.

The orientation of porous silicon influences of performance of silicon solar cells [45]. (p^+-n) silicon solar cells were prepared on base of n-type Si of (111) and (100) orientation (0.75 Ω cm). p^+-type doping was achieved by boron diffusion. Porous silicon layer on p^+-silicon was fabricated by electrochemical anodization in 1HF: 3C_2H_5OH solution at a current density of 50 mA/cm^2 for 20 min. The surface reflectance of porous silicon on (100) Si is significantly less (8 %) than that for (111) Si (16%). The photovoltaic parameters for (100) Si and (111) Si-based solar cells presented in Table 3show that the conversion efficiency of (100) Si-based cells is larger (15.42%) than that of (111) Si-based cells (12.4%). The observed results authors explain formation higher pyramids on (100) Si surface compared to (111) Si surface that results in lower reflectance.

It is to be noted that for thick porous silicon layer when it thickness is larger than that of p^+*(or* n^+*)*-emitter, non-uniformity of p^+-n (or n^+-p) junction profile can make worse the parameters of cells. Influence of different PS layer thickness on photovoltaic properties of n^+-p silicon solar cells showed that both external and internal quantum efficiencies are improved as far as PS/Si interface does not overpass the n^+-p junction [46]. It can mean that when the thickness of PS layer is larger than that ofn^+emitter region, i.e. the space-charge region locates in PS layer, the surface defects of PS layer stimulate the increase the recombination rate of photocarriers.

The comparative investigations of (n^+-p) Si (111) solar cells with single layer ARC (SiO_2), double layers ARC (ZnO/TiO_2) and the porous silicon layer are carried out in [47]. The porous silicon layers were formed on phosphorus-doped n^+-layer of solar cells in 1HF: 3C_2H_5OH solution at current density 60 mA/cm^2 for 30 min. The best photovoltaic characteristics showed the silicon solar cells with porous layer (see Table 3). From SEM micrograph presented in the work it is seen that thickness of porous silicon is significantly larger than the depth of p^+-n junction in silicon. It can make worse the characteristics of the p^+-n cells.

Above we considered the results of studies on silicon solar cells with porous layer formed only on front side of cell. Interesting data for silicon solar cells with porous layers prepared on both sides of the (111) n-type Si wafers (of 0.75 Ω cm resistivity and of 283 μm thickness) are given in [48]. The porous silicon layers on the polished front side and the unpolished backside were obtained by electrochemical process in 1HF: 4 C_2H_5OH solution under 60 mA/cm^2 for 15 min for each side (before boron diffusion and metal contact deposition). SEM images showed that the porous surface formed on the front polished side has discrete pores, whereas the porous silicon surface created on unpolished backside is shaped in small pores which are attributed to an increase in surface roughness. The reflectance of the PS front and back sides in the spectral range 400-1000 nm are about 16% and 6% respectively. As is seen from Table 3 the efficiency of (p$^+$-n) PS/nSi/PS solar cell with PS layer on both sides of Si is increased to 12.75% compared to solar cell with PS layer on the unpolished side (7.4%).

The formation of PS layer during the normal cleaning sequence at the beginning of the solar cell process was carried out in [49]. Porous silicon layers on p-type silicon wafer prepared by stain etching in HF:HNO_3 showed reflectance as low as 9 %. Photovoltaic characteristics of the best porous silicon cells fabricated by screen-printed technique are given in Table 3. The high values of photovoltaic parameters of (n^+-p) PS/pSi solar cells are attributed to lower reflectance and passivation properties of PS layer.

Thin porous silicon layer formed on dendritic web and string ribbon silicon by chemical etching in $HF:HNO_3:H_2O$ solution for a few seconds increases the lifetime of the minority carriers by a factor 3.3 [50]. The additional heat treatment (860°C, 2 min) or simultaneous phosphorus and aluminum diffusion after PS layer formation results in lifetime enhancement by a factor 8.3 or 5.8 respectively. In opinion of authors [50] porous silicon induced lifetime enhancement can be caused by the gettering properties of PS layer. Photovoltaic characteristics of (n^+-p) silicon solar cells with porous silicon layer and SiN antireflection coating are given in Table 3. It is seen that both type solar cells have almost same parameters.

The improvement of photovoltaic parameters of multicrystalline (n^+-p) Si solar cells with porous silicon layer is observed in [51]. Formation of porous silicon layer on textured surface of p-type mc-Si (1Ω cm) was performed by two-step chemical etching of silicon before phosphorus diffusion. It is shown that macroporous silicon layer formed before phosphorus diffusion results in texturisation of mc-Si with the larger angle than that obtained classical texturisation based in KOH. The surface reflectance of mc-Si with PS layer with phosphorus glass drops up to 5% compared to that for mc-Si with phosphorus silica glass but without PS layer (16%). The main parameters of (n^+-p) mc-Si based solar cells (of 25 cm^2 size) without and with PS layer were presented in Table 3. Formation of PS layer on mc-Si allows enhance the efficiency up to 12.7% that is close the efficiency limit of the mc-Si solar cells with TiO_x antireflection coating. Additional etching of (n^+-p) mc-Si solar cell in $98HNO_3$:2HF solution for 300 s (after PS formation) improves its photovoltaic parameters [7] (seeTable 3).

Gettering behavior of PS layer on silicon leading to enhancement of the lifetime of the non-equilibrium minority carriers and improvement of silicon solar cell characteristics have been considered in [52-54]. The porous silicon containing a large number of small pores with diameter between 4-50 nm shows very large the surface area-to-volume ratio (about of 500 cm^2/m^3). The extremely large pore surfaces and their very chemical activity ensure possibilityof application related with gettering of impurities and defects. The high-temperature annealing of the chemical etched porous silicon surface can enhance the impurity diffusion into the porous silicon network and thereby acting as an efficient external gettering site. The oxidation rate of porous silicon due to its mesapore structure is significantly larger (by 10-20 times) than that for single-crystalline silicon surface [52]. Gettering by porous silicon consists of oxidizing the porous silicon layer in wet oxygen ambient condition followed by the removal oxide in a dilute HF solution. The lattice mismatch between porous layer and silicon substrate can stimulate the gettering of impurities and defects.

The gettering-induced enhancement in the minority carrier diffusion length in multicrystalline silicon (mc-Si) with porous silicon layer on its surface was studied in [53]. The PS/Si/Al structures prepared by formation of porous layer in front surface of silicon and by vacuum evaporation of aluminum layer (2 μm) on the other side of sample were exposed to the wet oxidation of the porous silicon (950°C). Measurements showed that the minority carrier diffusion length for Si in such structure (about 190 nm) is larger than that for reference Si sample (about 100 nm). Co-gettering effect for PS/Si/Al consists in out-diffusion of impurities throughout the porous silicon layer and the aluminum. For {phosphorus layer/silicon/aluminum layer} structures (without porous layer) usually used for fabrication of silicon solar cell influence of co-gettering on the diffusion length of minority carriers is less (about 100 nm).

The PS/Si/Si structures prepared by forming porous silicon layers on both sides of Si wafer and then coated with phosphorus dopant ($POCl_3$) and heat treated at 900°C for 90 min discovered the improvement of the electrical and recombination characteristics of silicon [54]. Diffusion of phosphorus into PS layer is accompanied by gettering of eventual impurities towards the phosphorus doped PS layer. As a result of removal of impurities, increasing the mobility of the majority carriers and the diffusion length of minority carriers are observed. Removal of eventual impurities and defects away from the device active regions allowed to improve output characteristics of silicon solar cells.

Above, the results of influence of the porous silicon as an active element in thick crystalline silicon solar cells have been considered. It is shown the porous silicon as antireflection coating significantly improves the performance of silicon solar cells. In last year's porous silicon layer were also used as sacrificial layer for fabrication of thin-film silicon solar cells. As known the monocrystalline silicon which intensively are used in commercial solar cell applications is high cost material. The reduction in the amount of high-quality expensive silicon material per solar cell is one of ways of lowering the cell coat. At present one of the technologies developed to fulfill the aim of reduction the cost of silicon solar cell is porous silicon (PSI)-transfer process [55]. The thickness of monocrystalline silicon solar cell prepared by PSI-process (about 5-50 μm) is significantly lower than that of crystalline silicon cell fabricated by standard technology (250-300 μm). The PSI-transfer process consists of four steps. *First,* double layers of porous silicon are fabricated by electrochemical etching on surface of monocrystalline silicon: the 1-2 μm-thickness low-porosity layer (20%) at the top and 350 nm-thickness high-porosity layer (50-60%) beneath. *Second,* thin monocrystalline silicon layer (about 5-50 μm thickness) epitaxial grows on top (low-porosity) layer. Low-porosity layer is of monocrystalline quality which allows the growth of a high quality epitaxial layer of silicon. *At third step* the epitaxial silicon layer is detached from silicon substrate through a high-porosity layer by lift-off technique and then it is transferred onto a foreign substrate. At final step thin-film silicon solar cell is fabricated by standard or other technology. Shortly after using of PSI-process cell efficiency gradually increased from 12.5% (in 1997) to 16.9% (in 2009). Today the record value of efficiency of 19.1% (S=4 cm^2, V_{oc}=650 mV, J_{sc}=37.8 mA/cm2, FF=0.78) for monocrystalline silicon solar cell of 43 μm thickness prepared by porous silicon transfer technique was demonstrated [56]. It should be noted that porous silicon transfer process is certainly complex for industrial application. Nevertheless it shows the high efficiency potential of the porous silicon transfer process.

CONCLUSIONS

The review of investigations of the use of the nanoporous silicon in silicon solar cell showed that an increase in the conversion efficiency (about of 25-30%) is achieved for PS/Si solar cell compared to a cell without a PS layer. At the same time, the performance of silicon solar cells with PS layer is more than that of silicon solar cells with conventional ARC. The lower value of effective reflectance (up to 3%) for nanoporous silicon layer that significantly reduces the optical losses is one of main reasons of improvement of performance of PS/Si solar cell. A wide-band gap nanoporous silicon (up to 1.9 eV) resulting in widening of the spectral region of photosensitivity of the cell to the ultraviolet part of solar spectrum may promote the increase the efficiency of silicon solar cells with PS layer. The internal electric field of porous silicon layer with variable band gap (due to decrease of porosity deep down) can stimulate an increase in short-circuit current (Figure 16).

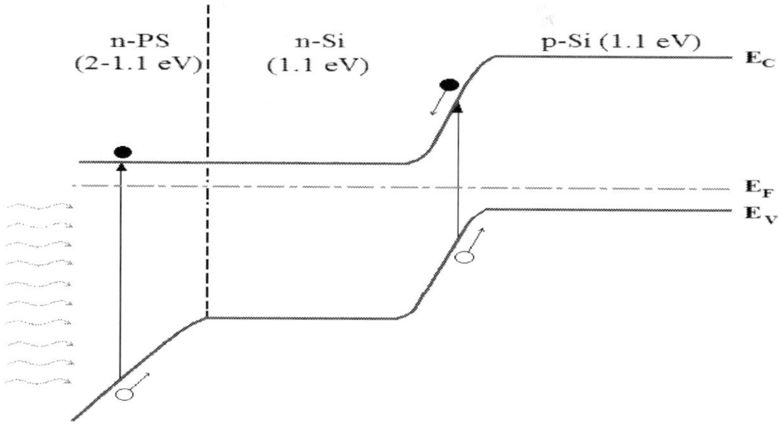

Figure 16: Energy band diagram of *n*PS/ (n^+-*p*) Si solar cell.

Additionally, the intensive photoluminescence in the red-orange region of the solar spectrum observed in porous silicon under blue-light excitation can increase the concentration of photo-exited carriers. It is necessary to take into account the passivation and gettering properties of Si-H and Si-O bonds on pore surfaces which can increase the lifetime of minority carriers.

Porous silicon, along with above advantages, in some cases e.g. due to its high resistivity can reduce the output parameters of PS/Si solar cells. Contribution of resistance (about 5×10^{-2} Ω) of thin porous silicon layer (about 100 nm) in total series resistance of the solar cell (about 0.5-1.0 Ω) is negligible and therefore it must not influence on parameters of cells. The resistance of thick PS layer (5-15 μm) can essentially increase the total series resistance and thereby it can reduce the photovoltaic parameters of silicon solar cell. However, as is seen from Table 3 reducing of parameters of PS/Si solar cells with thick PS layer was not observed. It tentatively can be explained that formation n^+or p^+-emitter layer by phosphorus or boron diffusion in PS is accompanied with decrease of layer resistance. Moreover, the nonflatprofile of n^{+}-p (or p^{+}-n) junction in PS/Si cells with thick PS layer (prepared before junction formation) does not result in reducing of efficiency (see Table 3). Explanation of this problem demands of further investigations.

Note should be taken that properties of PS layer and thereby photovoltaic characteristics of solar cell can change on running under illumination, heating etc. At present, as far as our knowledge goes, publications on temporal stability of PS-based solar cells are almost absent in literature.Works related with degradation phenomena in PS/Si solar cells is the matter of topical interest for further researches.

Judging by the results presented in this review and taking into account the simplicity of fabrication of porous silicon layer on silicon we can safely draw to the conclusion that the nanoporous silicon is a good candidate for use on preparation of low cost silicon solar cells with high efficiency. It gives hope for the industrial production of PS-based silicon solar cells.

REFERENCES

1. A. Goetzberger, C. Hebling, 2000 Photovoltaic Materials, Past, Present, Future Solar Energy Materials and Solar Cells 62 1 19
2. S. Sze, Physics of Semiconductor Devices New York: John Wiley & Sons 1990
3. L. Canham, 1997 Properties of Porous Silicon. London: IEE-Inspec.
4. O. Bisi, S. Ossicini, L. Pavesi, 2000 Porous Silicon: A Quantum Sponge Structure for Silicon Based Optoelectronic Surface Science Reports 38 1 126
5. A. Fahrenbruch, R. Bube, Fundamentals of Solar Cells, New York 1983
6. S. Gupta, M. Srivastaya, A. Gupta, Mathematical Formulation Comparative Analysis of Losses in Solar Cells. http://www.tutorialspoint.com/white-papers.
7. M. Lipinski, P. Panek, R. Ciach, The Industrial Technology of Crystalline Silicon Solar Cells J. Optoelectronics and Advanced Materials 2003 5 1365 1371

8. D. Neuhaus, A. Munzer, Industrial Silicon Wafer Solar Cells Advances in Electronics 2008 4 2 15
9. H. J. Moller, 1993 Semiconductor for Solar Cells. Boston: Artech House
10. M. Green, 1999 Limiting Efficiency of Bulk and Thin-film Silicon Solar Cells in Presence of Surface Recombination Progress in Photovoltaic 1999 7 3 327 330
11. I. Neizvestny, 2011 Nano-technologies in Solar Cells International Association of Academies of Sciences, Scientific Council on New Materials, conference proceedings October 5-7, 2011, Kiev
12. M. Born, E. Wolf, 1970 Principles of Optics. London: Pergamon Press.
13. A. Goodman, Optical Interference Method for the Approximate Determination of Refractive Index and Thickness of a Transparent Layer Applied Optics 1978 17 17 2779 2789
14. B. Sopori, R. Pryor, 1983 Design of Antireflection Coatings for Textured Silicon Solar Cells Solar Cells 1983 8 249
15. R. Jayakrishnan, Dielectric Coating Agents for Passivation and Anti-reflection Photovoltaic International 2009 6 83 86
16. Z. Chen, P. Sana, J. Salami, 1993 A Novel and Effective PECVD SiO2/SiN Antireflection Coating for Si Solar Cells IEE Trans. Electron Device 40 6 1161 1165
17. A. Mahdjoub, Graded Refractive Index Antireflection Coatings Based on Silicon and Titanium Oxides Semiconductor Physics, Quantum Electronics and Optoelectronics 2007 10 1 60 66
18. Z. Feng, R. Tsu, 1994 Porous Silicon. Singapore: World Scientific.
19. M. Beale, J. Uren, N. Chew, A. Cullis, An Experimental and Theoretical Study of the Formation and Microstructure of Porous Silicon Journal of Crystal Growth 1985 73 622 636
20. L. Remache, A. Mahdjoub, E. Fourmond, J. Dupuis, M. Lemiti, Design of Porous Silicon/PECVD SiOxAntireflection Coatings for Silicon Solar Cells Intern. Conference on Renewable Energies and Power Quality (ICREPQ-2010) Granada, Spain, 23-25 March 2010
21. T. D. Dzhafarov, B. Can, The Diffusion Distribution of Hydrogen and Oxygen in Porous Silicon Films Journal of Materials Science Letters 2000 19 287 289

22. G. B. Abdullaev, T. D. Dzhafarov, 1987 Atomic Diffusion in Semiconductor Structures. New York: Harwood Academic Press.
23. T. Dzhafarov, S. Aydin, Diffusion of Hydrogen in Porous Silicon-based Sensors Journal of Porous Media 2010 13 2 97 102
24. R. Cisneros, H. Pfeiffer, C. Wang, Oxygen Absorption in Free-Standing Porous silicon: A Structural, Optical and Kinetic Analysis Nanoscale Res. Lett. 2010 5 4 686 691
25. K. Barla, R. Herino, G. Bomchil, J. Pfister, A. Freund, Determination of Lattice Parameters and Elastic Properties of Porous Silicon by X-ray Diffraction Journal of Crystal Growth 1984 8 727 732
26. K. Barla, R. Herino, G. Bomchil, J. Pfister, J. Baruchel, X-ray Topographic Characterization of Porous Silicon Layers Journal of Crystal Growth 1984 68 721 726
27. D. Bellet, G. X. Dolino, X-Ray Diffraction Studies of Porous Silicon Thin Solid Films 1996 276 1 778 782
28. E. Perez, 2010 Design, Fabrication and Characterization of Porous Silicon Multilayer Optical Devices. Ph.D. thesis, Universitet Rovirai Virdali.
29. T. Dzhafarov, S. Aydin, 2009 Nano-Porous Silicon for Gas Sensors and Fuel Cells Applications Journal of Qafqaz University 25 20 35
30. T. Dzhafatov, C. Oruc, S. Aydin, Humidity-Voltaic Characteristics of Au-Porous Silicon Interfaces Journal of Physics D: Applied Physics 2004 37 404 408
31. C. Pickering, M. Beale, D. Robbins, R. Pearson, R. Greef, Optical Studies of the Structure of Porous Silicon Films Formed in p-type Degenerate and Non-degenerate Silicon J. Phys. C 1984 17 6535 6552
32. L. Schirone, G. Sotgiu, M. Montecchi, A. Parisini, Porous Silicon in High Efficiency Large Area Solar Cells Conference proceedings, Progr.14 European PV Solar Energy Conf. Barcelona, Spain 1479 1482 1997
33. S. Strehlke, S. Bastide, J. Guillet, J. Levy-Clement, Design of Porous Silicon Antireflection Coatings for Silicon Solar Cells Materials Science and Engineering 2000 B69-70 81 86

34. M. Lipinski, P. Panek, E. Beilanska, J. Weglowska, H. Czternastek, 2000 Influence of Porous Silicon on Parameters of Silicon Solar Cells Opto-Electronics Review 8 4 418 420

35. O. Weiying, Z. Lei, D. Hongwei, Z. Jun, W. Wenjing, Optical and Electrical Properties of Porous Silicon Layer Formed on the Textured Surface by Electrochemical Etching Journal of Electronics 2011 32 5 1 5

36. J. Kim, Formation of a Porous Silicon Antireflection Layer for a Silicon Solar Cell Journal of Korean Physical Society 2007 50 4 1168 1171

37. T. Dzhafarov, S. Aslanov, S. Ragimov, M. Sadigov, S. Aydin Yuksel, 2012 Effect of Nanoporous Silicon Coating on Silicon Solar Cell Performance Vacuum 86 12 1875 1879

38. M. Green, Limiting Efficiency of Bulk and Thin-film Silicon Solar Cells in the Presence of Surface Recombination Progress in Photovoltaic 1999 7 327 338

39. L. Canham, Silicon Quantum Wire Array Fabrication by Electrochemical and Chemical Dissolution of Wafers Appl. Phys. Lett. 1990 57 1046

40. M. Saadoun, H. Ezzaaoia, B. Bassais, M. Boujmil, R. Bennaceur, Formation of Porous Silicon for Large-area Silicon Solar cells: A New Method Solar Energy Materials and Solar Cells 1999 59 377 385

41. V. Yerokhov, I. Melnik, A. Tsisaruk, I. Semochko, Porous Silicon in Solar Cell Structures Opto-Electronics Review 2000 8 4 414 417

42. P. Panek, M. Lipinski, H. Czternastek, Porous Silicon Layer as Antireflection Coating in Solar Cells Opto-Electronics Review 2000 8 1 57 59

43. R. Chaoui, A. Messaoud, 2007 Screen-Printed Solar Cells with Simultaneous Formation of Porous Silicon Selective Emitter and Antireflection Coating Desalination 209 118 121

44. L. Remache, A. Mahdjoub, E. Fourmond, J. Dupuis, M. Lemiti, 2010 Design of porous silicon/PECVD SiOx antireflection coatings for silicon solar cells International Conference on Renewable Energies and Power Quality (ICREPQ-2010) 23-25 March, Granada

45. W. Aziz, A. Ramizy, K. Ibrahim, K. Omar, Z. Hassan, Effect the Orientation of Porous Silicon on Solar Cell Performance. Optoelectronics and Advances Materials 2009 3 12 1368 1370
46. S. Bastide, A. Albu-Yaron, S. Strehlke, C. Levy-Clement, Formation and Characterization of Porous Silicon Layers for Application in Multicrystalline Silicon Solar Cells 1999 57 393 417
47. A. Ramizy, W. Aziz, Z. Hasan, K. Omar, K. Ibragim, The Effect of Porosity on the Properties of Silicon Solar Cells Microelectronic International 2000 27 2 117 120
48. A. Ramizy, Z. Hassan, K. Omar, Y. Al-Douri, M. Mahdi, New Optical Features to Enhance Solar Cell Performance Based on Porous Silicon Surfaces Applied Surface Science 2001 257 6112 6117
49. R. Ludemann, B. Damiani, A. Rohatgi, 2000 Novel Processing of Solar Cells with Porous Silicon Texturing In: Proceedings of the 28-th IEEE Photovoltaic Specialists Conference 15-22 September, Anchorage, Alaska 299 301
50. M. Hilali, B. Damiani, A. Rohatgi, Lifetime Enhancement During Processing of Porous Silicon Cells. http://smartech.gatech.edu/jspui/bitstream.
51. P. Panek, Effect of Macroporous Silicon Layer on Opto-Electrical Parameters of Multicrystalline Silicon Solar Cells Opto-Electronics Review 2004 12 1 45 48
52. K. Hirscheman, L. Tsybeskov, S. Duttagupta, P. Fauchet, 1996 Silicon-based Visible Light-Emitting Devices Integrated into Microelectronic Circuits Nature 384 338 340
53. P. Vinod, Porous Silicon and Aluminum Co-gettering Experiment in p-Type Multicrystalline Silicon Substrates Science and Technology of Advanced Materials 2007 8 231 236
54. N. Khedher, M. Hajji, A. Ben Jaballah, B. Quertani, 2005 Gettering of Impurities from Crystalline Silicon by Phosphorus Diffusion Using a Porous Silicon Layer Solar Energy Materials & Solar Cells 87 1-4 605 611
55. C. Solanski, R. Bilyalov, J. Poortmans, J. Nijs, R. Mertens, Porous Silicon Layer Transfer Processes for Solar Cells Solar Energy Materials and Solar Cells 2004 83 101 113

56. H. Petermann, D. Zielke, J. Schmidt, F. Haase, E. Rojas, R. Brendel, 2012 Efficient and 43μm-Thick Crystalline Si Solar Cell from Layer Transfer Using Porous Silicon Progress in Photovoltaic: Research and Application 20 4 1 5

Chapter 5

Optimization of Third Generation Nanostructured Silicon-Based Solar Cells

Foozieh Sohrabi[1], Arash Nikniazi[2], and Hossein Movla[2]

[1]School of engineering emerging technologies, University of Tabriz, Tabriz, Iran

[2]Faculty of Physics, University of Tabriz, Tabriz, Iran

INTRODUCTION

Recently, the demand of solar cells has rapidly been growing with an increasing social interest in photovoltaic energy. Improving the energy conversion efficiency of solar cells by developing the technology and concepts must be increasingly extended as one of the key components in our future global energy supplement, but, the main problem of photovoltaic modules are their rather high production and energy cost.

Third generation solar cell is an alternative type of the promising device, which aims to achieve high-efficiency devices with low cost in comparison with expensive first generation solar cells and low-efficiency second generation solar cells. One of the prominent types is Si-based third generation solar cells which benefit from thin film processes and abundant, nontoxic materials. To gain efficiencies more than Shockley and Queisser limit which states the theoretical upper limit of 30% for a standard solar cell and overcome the loss mechanisms in this generation, different methods have been proposed:

1. Utilization of materials or cell structures incorporating several band gaps:
   - Si-based multi-junction solar cells
2. Modification of the photonic energy distribution prior to absorption in a solar cell:
   - Photon energy down-conversion
   - Photon energy up-conversion
3. Reducing losses due to thermalization:
   - Hot carrier solar cells
   - Impact ionization solar cells

This chapter mainly brings out an overview of the optimization of the first strategy and briefly the second and third strategies accompanying nanostructures. Multijunction solar cells are stacks of individual solar cells with different energy threshold each absorbing a different band of the solar spectrum. Si-based tandems based on quantum dots (QDs) and quantum wires (SiNWs) allowing band gap engineering and their optimization methods influencing their optical and electrical properties such as suitable Si QDs and SiNWs fabrication methods in various matrixes, interconnection between QDs, optimized impurity doping, etc. will be discussed. Moreover, the effects of the spacing and the size of Si QDs and SiNWs and their efficient amounts considering the latest researches will be introduced.

Another important process is multiple exciton generation (MEG) in QDs due to the current scientific interest in efficient formation of more than one photoinduced electron-hole pair upon the absorption of a single photon for improving solar devices.

Afterwards, the structural and superficial effects on the optimization of Si-based third generation solar cells like light concentration and use of forming gas will be presented.

Finally, the outlook concerning the mentioned methods will be suggested.

PRINCIPLE OF THIRD GENERATION SOLAR CELLS BASED ON SILICON

The main aim of third generation solar cell is obtaining high efficiency. To achieve such efficiency improvements, devices aim to circumvent the Shockley-Queisser limit for single-bandgap devices that limits efficiencies to either 31% or 41%, depending on concentration ratio (Fig. 1).

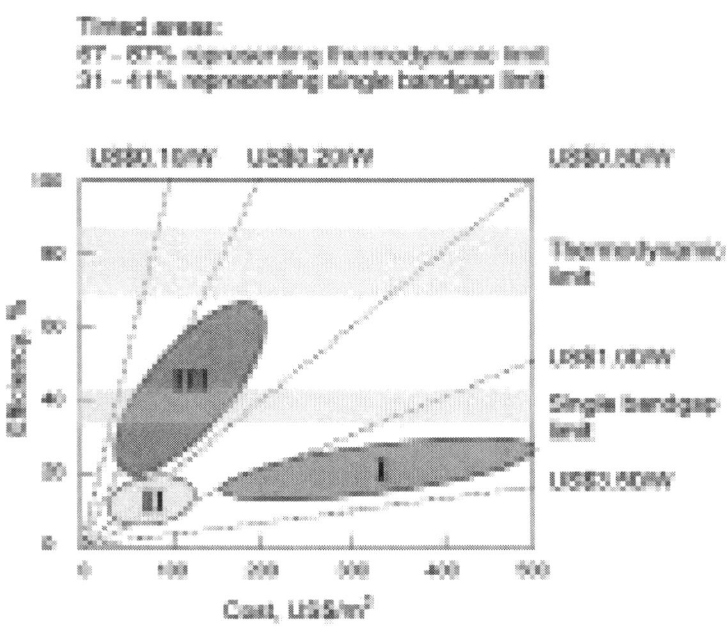

Figure 1: Efficiency and cost projections for first- (I), second- (II), and third-generation (III) PV technologies (wafer-based, thin films, and advanced thin films, respectively) [2].

The two most important power-loss mechanisms in single band gap cells are the inability to absorb photons with energy less than the bandgap (1 in Fig. 2) and thermalization of photon energies exceeding the bandgap in which the excess energy is lost as heat because the electron (and hole) relaxes to the conduction (and valence) band edge. The amounts of the losses are around 23% and 33% of the incoming solar energies, respectively (2 in Fig. 2) (1). Eventually, these two mechanisms alone cause the loss of about half of the incident solar energy in solar cell conversion to electricity. Other losses are junction loss, contact loss, and recombination loss which is shown in Fig. 2[1].

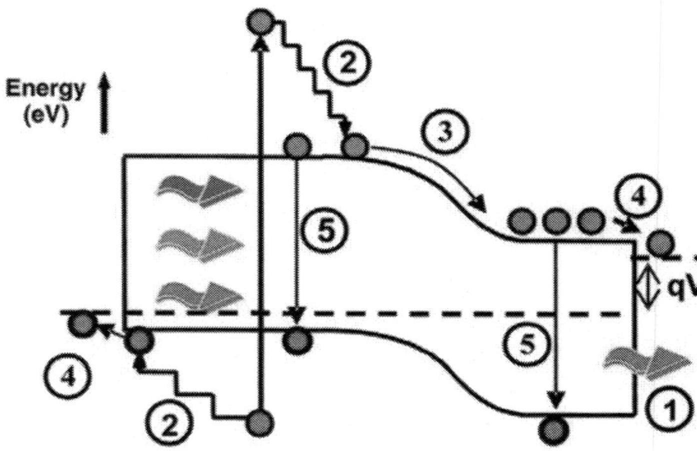

Figure 2: Loss processes in a standard solar cell: (1) non absorption of below bandgap photons; (2) lattice thermalization loss; (3) and (4) junction and contact voltage losses; (5) recombination loss (radiative recombination is unavoidable) [2].

Three families of approaches have been proposed for applying multiple energy levels:

1. increasing the number of energy levels;
2. multiple carrier pair generation per high energy photon or single carrier pair generation with multiple low energy photons; and
3. capturing carriers before thermalization.

Of these, tandem cells, an implementation of strategy (a), are the only ones that have, as yet, been realized with efficiencies exceeding the Shockley-Queisser limit [2].

In this chapter, firstly the concept of tandem solar cell or multijunction solar cell will be discussed and then Si nanostructured tandems will be explained precisely. However, amorphous silicon (a-Si) tandems will not be investigated in this chapter due to their lower efficiency in comparison with Si nanostructured tandem solar cells.

Multiple- Junction Solar Cell

One of the promising methods to enhance the efficiency of solar cells is to use a stack of solar cells, in which each cell has a band gap that is optimized for the absorption of a certain spectral region [3]. In fact, by stack layers, the number of energy levels is increased. This method was suggested for the first time by Jackson in 1955.

Solar cells consisting of p-n junctions in different semiconductor materials of increasing bandgap are placed on top of each other, such that the highest bandgap intercepts the sunlight first [2].

The importance of multijunction solar cell is that both spectrum splitting and photon selectivity are automatically achieved by the stacking arrangement.

To achieve the highest efficiency from the overall tandem device, the power from each cell in the stack must be optimized. This is done by choosing appropriate bandgaps, thicknesses, junction depths, and doping characteristics, such that the incident solar spectrum is split between the cells most effectively. Moreover two configurations are used for extracting electrical power from the device effectively which are reviewed by Conibeer: either a 'mechanically stacked' cell, in which each cell in the stack is treated as a separate device with two terminals for each; or an 'in-series' cell with each cell in the stack connected in series, such that the overall cell has just two terminals on the front and back of the whole stack. For a fixed solar spectrum and an optimal design, these two configurations give the same efficiency. But for a real, variable spectrum, the mechanically stacked design gives greater flexibility because of the ability to optimize the I-V curve of each cell externally and then connect them in an external circuit.

The reduced flexibility of just optimizing the I-V curve for the whole stack, because the same current must flow through each cell, makes the in series design more sensitive to spectral variations. Furthermore, they become increasingly spectrally sensitive as the number of bandgaps increases. For space-based cells this is not a great problem because of the constant spectrum, but for cells designed for terrestrial use, it is significant because of the variability of the terrestrial solar spectrum. This is particularly the case at the beginning and end of the day when the spectrum is significantly red shifted by the thickness of the atmosphere. Nonetheless, the much greater ease of fabrication of in-series devices makes them the design of choice for most current devices [2].

The efficiency depends on the number of subcells [1]. The efficiency limit for a single p-n junction cell is 29%, but this increases to 42.5% and 47.5% for 2-cell and 3-cell tandem solar cells, respectively. However, these values are a little bit more for concentrated light.

For example, the radiative efficiency of bulk silicon (Si) solar cells under the AM1.5G spectrum is limited theoretically to 29% due to the incomplete utilization of high energy photons and transmission of photons with less energy than the Si band gap [3]. But, the theoretical efficiency of tandem solar cells with a bulk Si bottom cell increases to 42.5 % when one additional solar cell with 1.8 eV band gap is used and to 47.5 % with two further solar cells with band gaps of 1.5 and 2 eV placed on top of the bulk Si cell.

Si Nanostructure Tandems

Silicon is not suitable for optoelectronic applications because of its indirect bandgap and poor light emission properties. However, silicon bandgap tuning above bulk silicon bandgap (1.12eV) is possible in the nanometer regime (sizes less than 10nm) enabling a revolutionary control over its properties [4].Therefore, use of nanostructures in tandem solar cells can create bandgap engineering besides improving the efficiency. Improved optical and electrical properties of silicon can be found in different forms of silicon, for example, porous silicon, silicon superlattices and Si-QD embedded in dielectric [4].

In silicon based tandem solar cells, this bandgap engineering can be done using either quantum wells (QWs) or quantum dots (QDs) of Si sandwiched between layers of a dielectric based on Si compounds such as SiO_2, Si_3N_4, SiON or SiC which taking advantages of the widening of absorption spectrum in the UV range [5]. As a whole, Si nanotechnology is the best choice to improve the metastabilities and to increase the quantum efficiency [6].

By restricting the dimensions of silicon to less than Bohr radius of bulk crystalline silicon (almost 5 nm), quantum confinement causes its effective bandgap to increase. If these dots are close together, carriers can tunnel between them to produce QD superlattices. Such super lattices can then be used as the higher bandgap materials in a tandem cell [1]. In fact, the idea is to add one or more layers of nano-structured materials on the top of a solar cell for which the optical absorption covers different domains of the solar spectrum (Fig. 3 is an example of "all silicon" tandem solar cell).

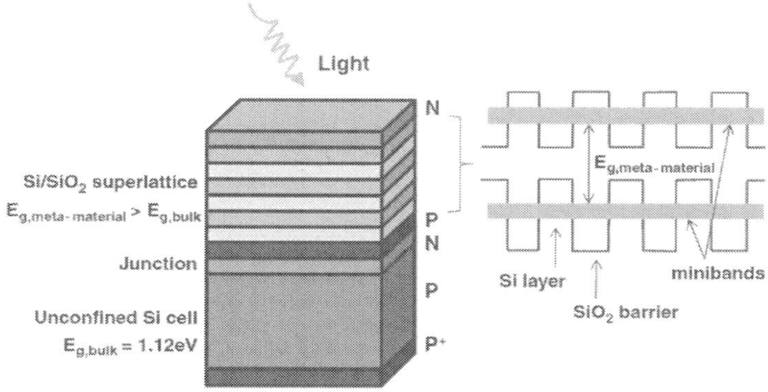

Figure 3: Schematics of an "all silicon" tandem solar cell with a top cell based on a nanostructured meta-material stacked on an unconfined Si cell [7].

All tandem solar cells offer the advantages of using silicon which is an abundant material, stable, non-toxic and capable to diversify in order to obtain both a medium bandgap material (~1 eV) and high a bandgap material (~1.7 eV) [7]. It should be mentioned that combining two tandem cell bandgaps (1.12 eV and 1.7 eV) achieve a conversion efficiency factor up to 42%.Another significant advantage of Si is its well developed technology in the world which paves the way for experimental and optimization studies of tandem solar cells. Moreover, strong optical absorption and high photocurrent have been found in nc-Si films and attributed to the enhancement of the optical absorption cross section and good carrier conductivity in the nanometer grains [8].

An approach to prepare silicon quantum dot super lattices by depositing alternating layers of stoichiometric oxide followed by silicon-rich oxide also appears promising in a potentially low cost process, with the control of dot diameter and one spatial coordinate [9].In detail, these layers are grown by thin-film sputtering or CVD processes followed by a high-temperature anneal to crystallize the Si QWs/QDs. The matrix remains amorphous, thus avoiding some of the problems of lattice mismatch [2]. For sufficiently close spacing of QWs or QDs, a true miniband is formed creating an effectively larger bandgap. For QDs of 2 nm (QWs of 1 nm), an effective bandgap of 1.7 eV results – ideal for a tandem cell element on top of Si [2].

Because of the charge carrier confinement in Si quantum dots it is possible to adjust the band gap by a control of the Si quantum dot size [3].

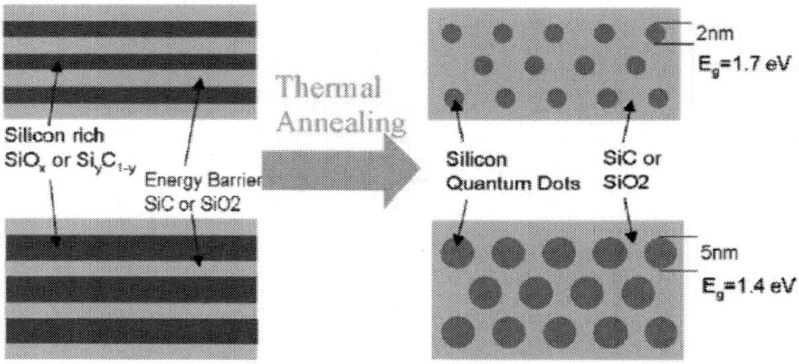

Figure 4: Schematic of the procedure to achieve size control of Si NC in Si based dielectric matrices. Layers with silicon excess are deposited alternately between stoichiometric layers. The stoichiometric layers act as a diffusion barrier for the silicon atoms and therefore limit the growth of silicon nano-crystals during the annealing step [3].

Generally, there are two ways for observing and estimating the size [1]:

1. The dot size of the Si QDs can be evidenced by high-resolution transmission electron microscopy (HRTEM). We can clearly see black dots due to contrast difference between Si and SiO_2 in Fig. 5.
2. Raman spectroscopy can also be used to estimate the dot size. (Fig. 6.)

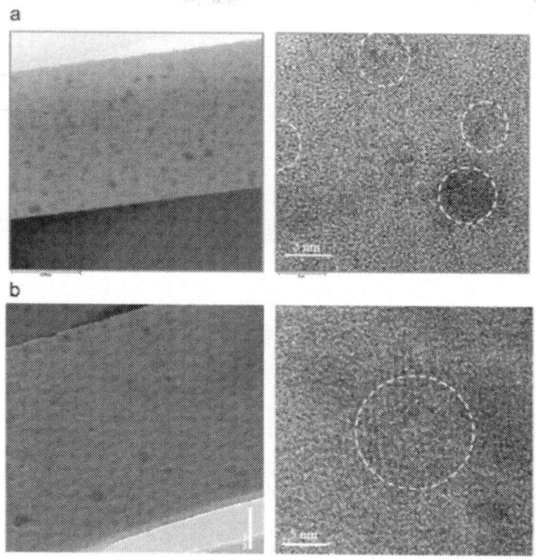

Figure 5: Transmission electron microscopy (TEM) images of Si quantum dots in SiO_2 matrix with low-magnification and high-resolution lattice images for (a) 5 nm Si QDs and (b) 871 nm Si QDs[1].

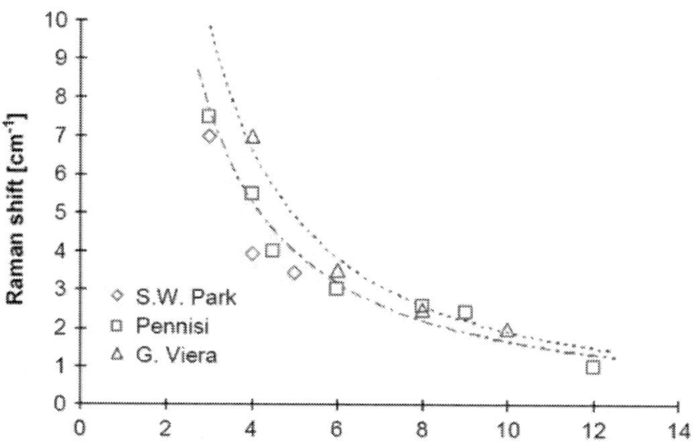

Figure 6: Raman peaks shifts to lower energy for Si QDs with 3,4, and5 nm. Reference data are adapted from Pennisi and co-workers [8] and Viera et al[1].

To realize all-silicon tandem solar cells, Park et al., fabricated phosphorus-doped Si QDs super lattice as an active layer on p-type crystalline Si (c-Si) substrate as shown in Fig. 7. The phosphorous doping in n-type Si QDs super lattice was realized by P_2O_5 co-sputtering during the deposition of silicon-rich oxide (SRO, Si and SiO_2 co-sputtering), which forms Si QDs upon high-temperature post-annealing. The n-type region typically includes 15 or 25 bi-layers formed by alternating deposition of P-doped QDs and SiO_2 [1].

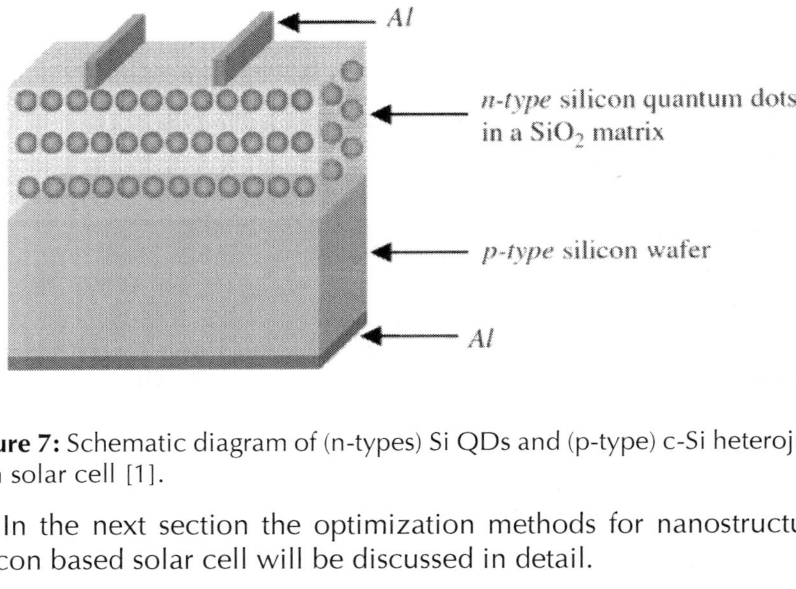

Figure 7: Schematic diagram of (n-types) Si QDs and (p-type) c-Si heterojunction solar cell [1].

In the next section the optimization methods for nanostructured silicon based solar cell will be discussed in detail.

OPTIMIZATION METHOD IN NANOSTRUCTURES

Silicon Quantum Dots Solar Cells

The main challenge for a nanostructure engineered material in a tandem cell is to achieve sufficient carrier mobility and hence a reasonable conductivity. For a nanostructure, this generally requires formation of a true superlattice with overlap of the wave function for adjacent quantum dots; which in turn requires either close spacing between

QDs or a low barrier height. Moreover, the quantum confinement, achieved by restricting at least one dimension of silicon less to the Bohr radius of the bulk crystalline silicon (around 5 nm), causes the effective bandgap to increase [10] which also results in increased absorption. The strongest quantum confinement effect is obtained if the silicon is constrained in all three dimensions, as in quantum dots, such that the same increase in effective bandgap can be achieved with a much less stringent size constraint [10]. Different technological approaches allowing formation of Si QDs. Generally, perfect (ideal) QD arrays can have the following characteristics [11]:

- Absolutely accurate positioning and control for nucleation site of individual QD;
- Uniformity of size, shape and composition;
- Large-area (~cm^2), long-range ordering QDs;
- The ability to control the QDs size;
- The ability to achieve both ultra-high dense QD arrays and sparse QD arrays

In this part we will discuss on optimum properties of Si QDs that include size, spacing and dielectric matrix of Si QDs which also have great influences on the band structure [9].

Optimum Size of SI QDS

A control of the Si nanocrystal size allows the adjustment of essential material parameters such as bandgap and oscillator strengths due to size quantization effects [3]. Experimental results have shown that the size of the QDs can be quite well controlled by selecting an appropriate thickness for the SRO layer and the density of the dots can be varied by the composition of the SRO layer. In detail, the size and crystallization of the Si nanocrystals are dependent on a number of factors, including the annealing method and the barrier thickness [12].

In 2006, Gavin Conibeer et al., at the University of NSW, used the energy confinement of silicon based quantum dot nanostructures to engineer wide band gap materials to be used as upper cell elements in Si based tandem cells. HRTEM data shows Si nanocrystal formation in oxide and nitride matrixes with a controlled nanocrystal size, grown by layered reactive sputtering and layered PECVD [13].

The data shown in Fig. 8 are measured from HRTEM images for samples at several deposition times. There is a sharp decrease in the nanocrystal size distribution on reduction in layer thickness from 4.7 to 3.5 nm. This indicates a transition from a bulk diffusion mechanism of Si atoms during precipitation to a constrained two dimensional diffusion regime, such that the nanocrystal size is defined by the layer thickness [13].

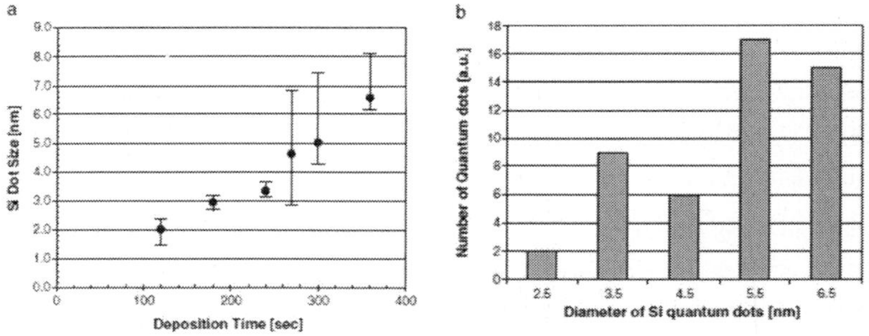

Figure 8: a) Dependence of quantum dot size distribution on deposition time as measured by HRTEM (other sputtering parameters optimized). b) QD size distribution for deposition time of 280 s [13].

This is an important self-regulation effect which gives much greater uniformity in nanocrystal optoelectronic properties, at least in the growth direction, as indicated for photoluminescence (PL). PL results indicate quantum confined properties as evidenced by the increase in the photo-luminescent energy in PL experiments [13].

Fig. 9a shows an increase in PL energy as nanocrystal size decreases, thus demonstrating quantum confinement and hence formation of quantum dots. It also shows a dramatic increase in PL intensity on going from a dot diameter of 4.7 to 3.5 nm. This correlates well with the greatly increased uniformity in Si quantum dot size as the deposited layer goes from 4.7 to 3.5 nm, as shown in Fig.8aon change of diffusion mechanism (see above). The large increase in PL energy is due to the much greater signal at a given energy with good dot size uniformity. (The fact that this intensity drops again is discussed below.) An increase in PL intensity is also to be expected as dot size decreases because of the increase in spatial localization of electrons and holes that will increase the probability of recombination [13].

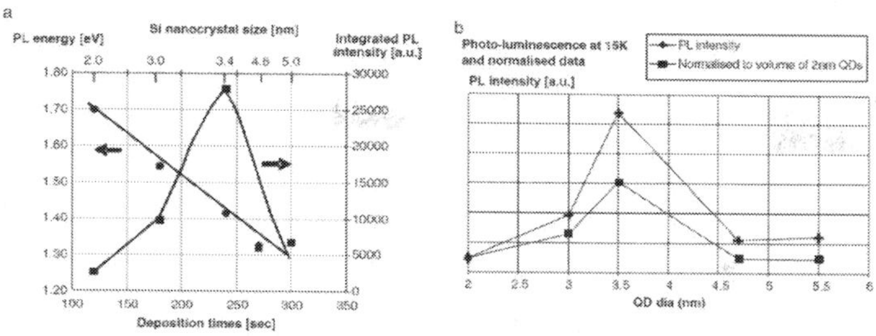

Figure 9: a) PL energy and integrated intensity (15 K) as a function of deposition time, showing quantum confined energy in silicon quantum dots. Deposition time is also calibrated for dot diameter by TEM. b) PL intensity data normalized for decreasing volume [13].

Eun-Chel Cho et al. in 2007 in Australia show that there is a large increase in PL intensity as the QD size decreases, which is consistent with the increase in radiative efficiency with the onset of pseudodirectbandgap behavior. The photoluminescence peaks from Si QDs in nitride are more blue-shifted than that of Si QD in oxide. Figure 10 shows the PL peak energies from Si QD dispersed in oxide and nitride. PL peak energies of Si QDs in oxide are less than 2.0 eV, while Si QDs in nitride have peak energies less than 3.0 eV [10].

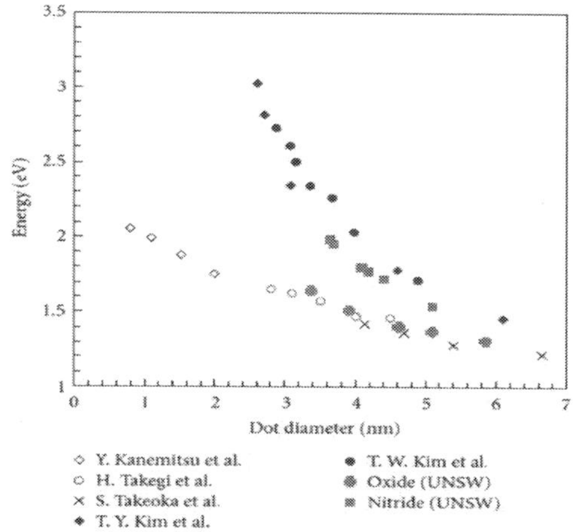

Figure 10: Energy gaps of three-dimensionally confined Si nanocrystals in SiO_2 and SiN_x (300°K) [10].

Puzder et al. in 2002 have explained that the main reason for the PL peak energy reduction in oxide matrix is the distortion of the local sp^3 network by double-bonded oxygen. However, Yang et al. in 2004 claimed that the reason for the stronger blue shift in nitride is better passivation of Si QDs by nitrogen atoms, eliminating the strain at the Si/Si_3N_4 interface nearly completely [10]. Generally, it can be concluded that the optimum size of Si QDs is 2-3nm.

Optimum Spacing of SI QDS

If QDs dots are spaced close enough and there is a significant overlap of the wave function, carriers can tunnel between them to produce a true quantum dot superlattice. Such a superlattice can then be used as the higher bandgap material in a tandem cell [10]. In other words, one common strategy to boost the performance of photovoltaic device structures is incorporating closely packed 3D QD array into device structures. When QDs in different size are formed into an ordered 3D array, there will be strong electron coupling between them so that excitons will have a longer life, facilitating the collection and transport of "hot carriers" to generate electricity at high voltage [11]. In addition, such an array makes it possible to create more than one electron-hole pair from a single absorbed photon, and thus enhance photocurrent, through the process of impact ionization [14]. This process happens when the energy of the photon is far greater than the semiconductor bandgap; while in bulk semiconductors the excess energy is simply dissipates away as heat, in QDs the charge carriers are confined within an infinitesimal volume, thereby increasing their interactions and enhancing the probability for multiple exciton generation [14].

The transport properties of the ensembles of disordered Si QDs in insulating matrix could be explained in terms of the percolation theory, which has already been successfully implemented to explain the transport processes in granular metals by Abeles et al. in 1975. Indeed, this theory describes the effect of the system's connectivity on its geometrical and physical properties [15].

To what concerns the ensemble of Si QDs, there can be distinguished five different structural-electrical regimes, such that in each of them we may expect a different transport mechanism to dominate. These regimes are:

1. Spherical QDs isolated by uniformly dispersed in insulating matrix
2. Some of the QDs starts to "touch" their neighbors named as Transition regime
3. Forming of clusters of "touching" QDs named as Intermediate regime
4. Clusters of regime 3 form a global continuous network named as Percolation transition regime
5. Well forming of percolation cluster of "touching" QDs and disappearance of geometrically non-"touching" QDs.

Fig.11 (a), (b) and (c) present typical examples of ensembles of Si QDs corresponding to regimes 1, 3 and 5, respectively. Usually there are narrow (no more than 0.5 nm wide) boundaries formed between the nanoparticles, which involves at least a different crystallographic orientation of the touching crystallites. In a literature the charge transfer process between such "touching" QDs was termed as "migration".

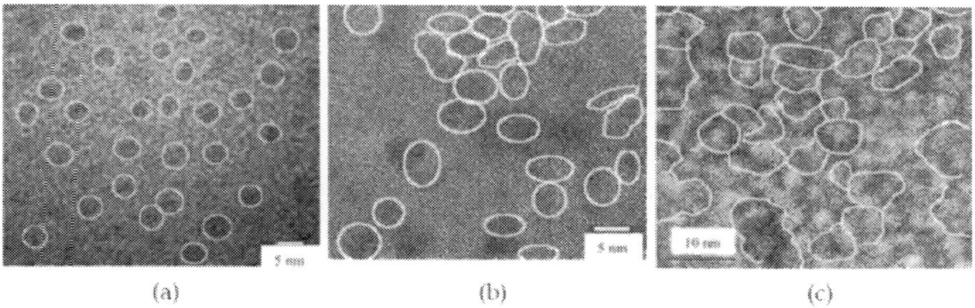

Figure 11: HRTEM images of the ensembles of Si QDs corresponding to different structural electrical regimes: (a) uniformly dispersed isolated spherical QDs (regime 1), (b) clusters of "touching" QDs (regime 3) and (c) percolation clusters of "touching" QDs (regime 5)[15].

The effect of the connectivity on the transport properties (dark and photoconductivity) of the ensembles of Si QDS is illustrated in Fig. 12. As one can see, the global picture of transport in Si QDs ensembles is reminiscent of that of granular metals, but the details are quite different. As long as Si QDs or clusters of Si QDs are small enough, they "keep" the carrier that resides in them and become charged when an excess charge carrier reaches them. Hence, the transport through the system can take place only if a corresponding charging (or Coulomb) energy is provided [15]. Balberg et al. in 2004 reported this topic for the samples with low number of Si QDs in the ensemble, which are characterized by the QDs of regime 1, the local conductivity is determined by the tunneling of charge carriers under Coulomb blockage between adjacent nanocrystallites similar to the case encountered in granular metals in the dielectric regime.

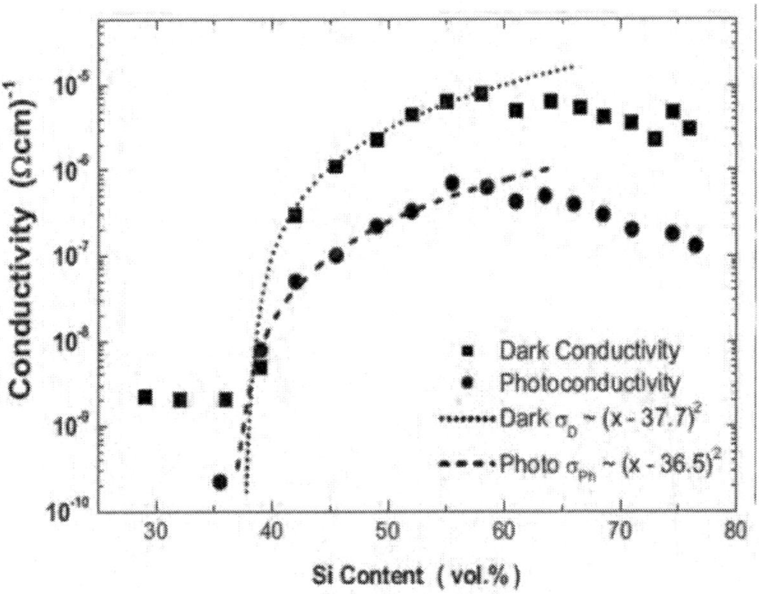

Figure 12: Dependence of the dark conductivity and the photoconductivity on the Si content [15].

With increasing of Si content (Figure 11b), the interparticle distance decreases and the tunneling-connected quantum dot clusters grow in size. The "delocalization" of the carrier from its confinement in the individual quantum dot to larger regions of the ensemble will take place, i.e., the charge carrier will belong to a cluster of QDs rather than to an individual QD. Correspondingly, this will also yield a decrease in the local charging energy in comparison with that of the isolated QD and the distance to which the charge carrier could wander will increase and as a consequence the conductivity of the ensemble will increase as well. The charge carrier transport in the case of regime 3 is thus determined by the intracluster migration and by the intercluster tunneling.

Finally we can conclude from Fig. 12 that the maximal possible conductivity is assured regime 5 and the highly percolating system of Si QDs will ensure the most favorable conditions for the electronic transport between the nanocrystals and the bandgap value in such structures can be adjusted in the large range covering the major part of the solar spectrum.

Also transport properties are expected to depend on the matrix in which the silicon quantum dots are embedded. As shown in Figure 13, different matrices produce different transport barriers between the Si dot and the matrix, with tunneling probability heavily dependent on the height of this barrier. Si_3N_4 and SiC give lower barriers than SiO_2 allowing larger dot spacing for a given tunneling current [13].

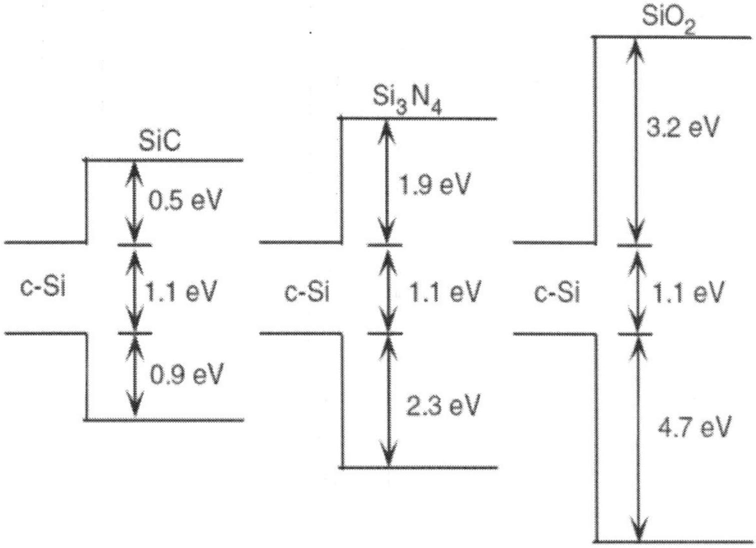

Figure 13: Bulk band alignments between crystalline silicon and its carbide, nitride and oxide [13].

The results suggest that dots in a SiO_2 matrix would have to be separated by no more than 1-2 nm of matrix, while they could be separated by more than 4 nm of SiC [10]. It is also found that the Bloch mobilities do not depend strongly on variations in the dot spacing but do depend strongly on dot size within the QD material [16].

Hence, transport between dots can be significantly increased by using alternative matrices with a lower barrier height, ΔE. The spacing of dots would have to be closest in the oxide, nitride and carbide, in that order. Similar deposition and quantum dot precipitation approaches should work for all [16].

Optimum Dielectric Matrix of SI QDS

In recent years it emerged that the Si QD interface with its surrounding dielectric matrix plays a decisive role in determining the optical absorption gap and the optical activity of the Si QD on both experimental and theoretical grounds [17].

Generally, as mentioned above, three types of dielectric matrices SiO_2, Si_3N_4, or SiC are used to form all tandem silicon solar cell.

It should be considered that lower barrier heights will give a greater tunneling probability between adjacent Si QDs and hence greater conductance [18]. Therefore, Si_3N_4 and SiC have greater conductance than SiO_2. In detail, SiC has the lowest barrier height among these dielectrics. However, the low barrier height also limits the minimum size of QDs to about 3nm or else the quantum-confined levels are likely to rise above the level of the barrier, which should be around 2.3 eV for amorphous SiC. In addition, although SiO_2 matrix has higher barrier height (3.2eV) comparatively, many attempts were made to fabricate Si-QD in Si-rich SiO_2 thick layers or superlattices since SiO_2 is a frequently used dielectric and compatible in microelectronics processes [4].

Si ion implantation into an oxide layer can be used to produce Si QDs at an irregular position with a relatively large size distribution. Si QDs can fabricate by solution synthesis, mechanical milling, and particle selection from porous silicon, but it is difficult to control the size uniformity of distributed QDs or an additional process to select the particle with the same size [15].

The material requirements for the dielectric layers are ease of thin-film growth and use of abundant nontoxic materials, hence it is most likely to be an oxide, nitride, or carbide of silicon. It is also necessary that carriers from the quantum dot layers have a high probability of tunneling through the dielectric layers [10]. It is worth noting that these devices must be thin to limit recombination due to their short diffusion lengths, which in turn means they must have high absorption coefficients [18].For layers of thickness less than about 4 nm, the precipitation enters a regime of 2D diffusion in which the dot size is accurately controlled by the layer thickness [13]. This is achieved by creating each dielectric layer with a thickness in the range of 1.5 to 2.5 nm for the case of oxide [10].

Si QD fabrication by various vacuum deposition techniques is preferable because of the greater potential of integration into conventional devices [10]. These include sputtering and plasma enhanced chemical vapor deposition (PECVD). The most successful and hence most commonly used technique is sputtering, because of its large amount of control over deposition material, deposition rate and abruptness of layers. This uses a new multi-target remote plasma sputtering machine with two independent RF power supplies as well as an additional DC power supply [16].

Si precipitation from a Si-rich layer, high temperature annealing of excess Si in an inert atmosphere is necessary to form Si nanocrystals with a few nm diameters, for example, Si QD precipitation in oxide, nitride, and carbide (Figure 14.a) and Equation (1) describes this Si precipitation mechanism [10]:

$$Si(O,N,C)_x \rightarrow \left(\frac{x}{2}\right) Si\left(O_2, N_{\frac{4}{3}}, C\right) + \left(1-\frac{x}{2}\right) S_i\,(1) \qquad (1)$$

It should be mentioned that in particular the amount of excess silicon in the Si-rich layer is an important parameter to study the nucleation of the QDs. If the Si concentration in the Si-rich layer film before annealing exceeds a certain limit, dots can merge together during the coalescence phase of the growth process, affecting the quantum confinement properties of the structures [25].

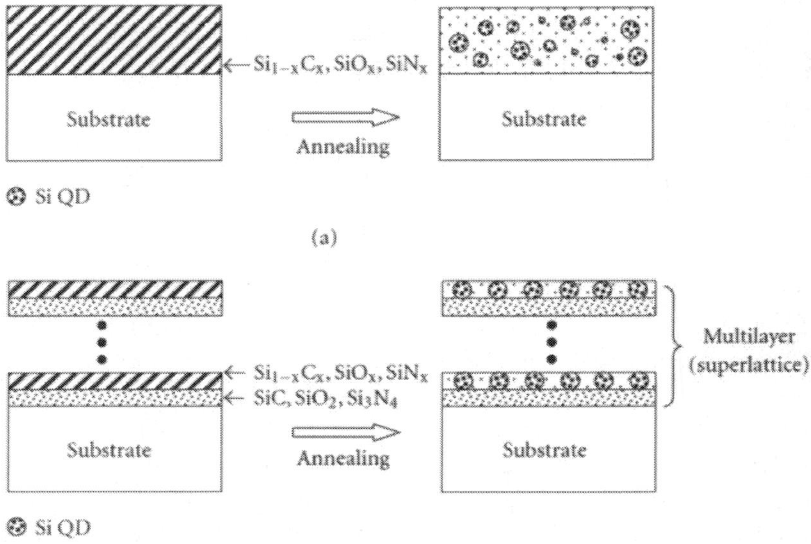

Figure 14: Si QDs from phase separation of a) a single silicon-rich precursor layer and b) a multilayer structure [10].

More accurate size control and a narrow size distribution are achieved by growth of a Si QD multilayer structure, which is fabricated by alternating layers of stoichiometric insulating materials and silicon-rich layers shown in Figure 14 b. Depending on the annealing conditions, silicon precipitates from the silicon-rich layers as approximately spherical QDs of a diameter close to the original layer thickness. Hence controls of the diameter and of one spatial coordinate of the dots are possible.

As mentioned above, SiO_2 is a frequently used dielectric and compatible in microelectronics processes. Therefore, Si-QD is generally fabricated in Si-rich SiO_2 thick layers or superlattices. Another dielectric option is SiN_x dielectrics. Due to low barrier height, highest Si-QD growth density in Si3N4 and less silicon requirement in Si_3N_4 during deposition, this dielectric is replaced instead of SiO_2. Moreover, the formation of Si-QD in SiN_x is preferable, because the formation of 3–7 nm Si-QD in Si-rich SiC film requires higher thermal budget (1100 1C) than Si-QD formation in a-Si/SiN_x layer structure that requires lower annealing temperature (800-850°C) [4].

Efficient photoluminescence (PL) has also been observed from So-rich SiN_x films, a single Si-QD layer sandwiched between two a-SiN_x layers and structured layers of Si-QD in a-SiN_x. In addition, for the same Si-QD size and PL excitation wave length, Si-QDs in SiN_x film show PL peak in shorter wavelengths (450-620 nm) than Si-QDs in SiO_2 film (650-950 nm) [4]. TEM images in Figure 8 show Si QDs interspersed in the oxide matrix [10].

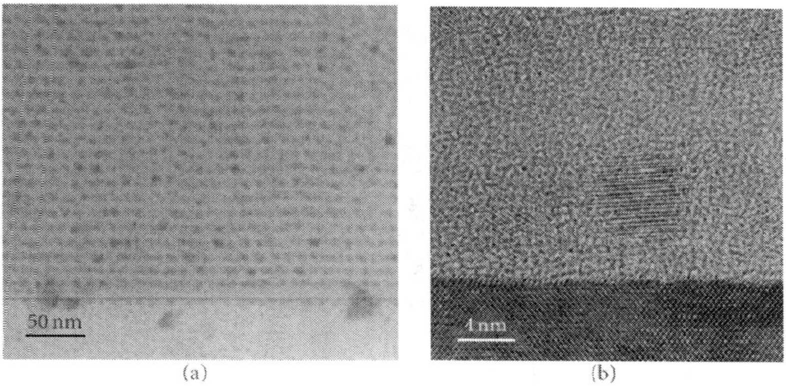

Figure 15: Transmission electron microscopy (TEM) images of Si QDs in SiO2 matrix with (a) low magnification and (b) HRTEM lattice image of Si quantum dots [10].

The bandgap of Si_3N_4 is significantly lower than that of SiO_2. Hence Si QDs in the nitride offer a lower barrier height and much increased carrier tunneling probability between quantum dots. (SiC offers a lower barrier still.) For this reason, Si QDs have explored transferring the technology in $SiO_2$to the growth of Si nanocrystals in silicon nitride by both sputtering and PECVD [10].

Layered Si QDs have also extended the in nitride technology to gas phase in situ deposition. Figure 16shows in situ Si QD dispersed in a nitride matrix. A stoichiometric Si_3N_4 layer and an in situ Si QD layer are alternately deposited on a Si substrate. This technique allows QDs to form during deposition without a post deposition annealing. This technique is a low temperature process and may be potentially beneficial to doping of Si QDs owing to high equilibrium temperature of the plasma and free of high temperature postannealing described in Figure 14 [16].

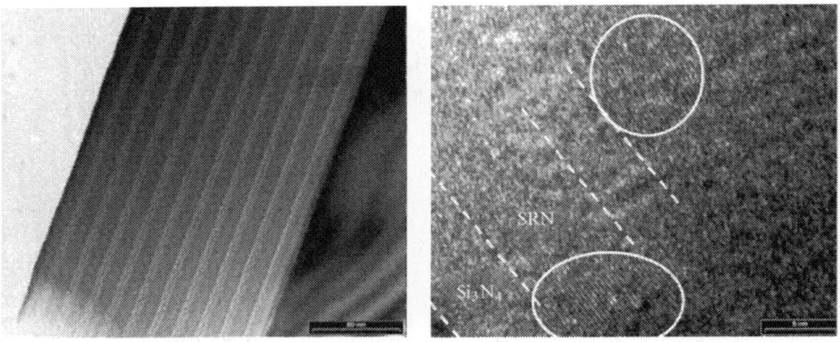

(a) (b)

Figure 16: Si QDs dispersed in a Si_3N_4 matrix fabricated by gas phase in situ deposition: a) low-magnification TEM and b) high-resolution TEM [10].

Another dielectric for growing Si-QD is SiC. Although the electron tunneling conductivity is higher in SiC compared to Si_3N_4 and SiO_2 due to the lower barrier height (0.5eV) of SiC; the formation of Si-QD in SiN_x is preferable, because the formation of 3–7 nm Si-QD in Si-rich SiC film requires higher thermal budget (1100°C) than Si-QD formation in a-Si/SiN_x layer structure that requires lower annealing temperature (800-850°C) [4].

After annealing above 1000°C, indicating formation of amorphous graphitic carbon is indicated. Hence the best data so far for Si nanocrystals in a SiC matrix are obtained for a $Si_{0.75}C_{0.25}$precursor composition [10]. $Si_{1\_x}C_x$/SiC multilayers have also been deposited by sputtering to give better control over the Si QD as with oxide and nitride matrices [18].

Raman, TEM and XRD spectra for a silicon-rich $Si_{0.75}C_{0.25}$precursor layer grown on a quartz substrate with subsequent annealing are shown in Figure 17. There is clear evidence for the formation of nano-crystalline Si at an annealing temperature greater than 1000°C. This is shown in the Raman peak at ~ 508 cm^{-1} (red shifted from 520 cm^{-1} due to a nanocrystalline folded Brillouin zone dispersion in k-space); TEM lattice fringe spacing consistent with {111} Si planes; and XRD peaks at $2\theta = 28.40$ with peak broadening indicating nanocrystal of 3-7nm (estimated using the Scherrer equation). It should be noted that here the nanocrystal size determined by TEM is slightly smaller than that determined by XRD[10].

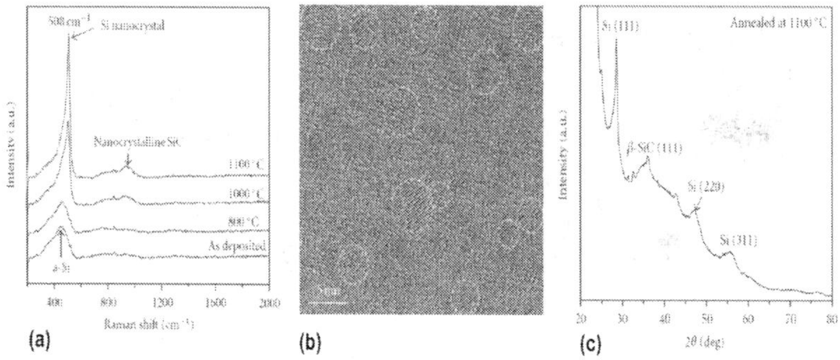

Figure 17: Silicon-rich SiC precursor layer: (a) Raman spectra for various annealing temperatures; Cross-sectional HRTEM image; and (c) X-ray diffraction[10].

Other Si and C concentrations were tried. As the concentration of C in SiC_x is increased to the nearly stoichiometric $Si_{0.495}C_{0.505}$, Raman evidence for the stretching vibration modes of Si-C and C-C bonds can be easily identified. With increasing annealing temperature, which shows increasing intensities of both TO and LO bands, the formation of crystalline SiC during annealing is indicated. In addition, there is a dramatic decrease in the intensity of Si-Si vibration modes indicating the formation of far fewer Si nanocrystals. There is also evidence for free carbon at ~1400 cm^{-1} in as deposited film, splitting into two bands at ~1360 (D band) and 1590 cm^{-1} (G band) [16].

As before, a multilayer approach was used to fabricate Si QDs in carbide of uniform size. A multilayer with stoichiometric SiC and silicon-rich $Si_{1-x}C_x$ precursor layer was fabricated, as shown in Figure 17.a, and annealed at a high temperature to selectively precipitate Si nanocrystals in a carbide matrix. However, the lattice fringes in HRTEM image correspond to β-SiC {111} crystalline planes (Figure 17.b). One possible reason for SiC QDs, instead of Si QDs, is that the C/Si ratio in layered structure has an increase, compared to the original design.

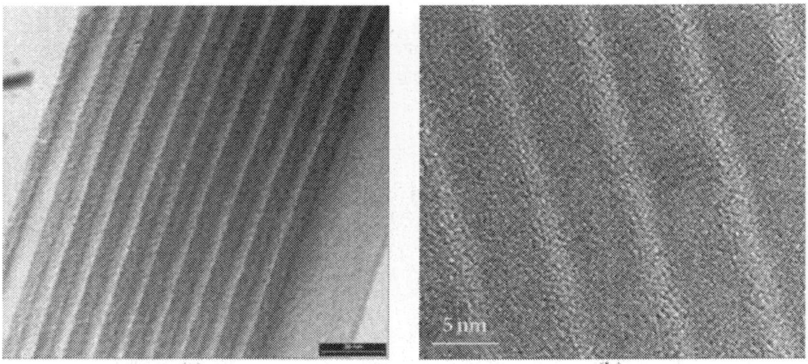

(a) (b)

Figure 18: TEM images of SiC/Silicon-rich SiC multilayer a) deposited and b) annealed at 1100°C for 20 minutes [10].

The decay length L_d is determined by the barrier height of the material in which the dot is embedded, which in the present case is either a silicon carbide, nitride, or oxide matrix [10]. In the simplest case, the decay length (L_d) is given by:

$$L_d = \frac{1}{\sqrt{2m^*(V_0 - E_n)}} \quad (2)$$

The latter expression holds when ΔE, the difference between the conduction band edge of the matrix and the confined energy level, is in eV.m* and m_0 are the effective mass and electron mass in the matrix material, respectively. V_0 is the corresponding band offset and E_n is the confined energy level in a quantum dot. Without considering the confined energy of QDs, energy difference ΔE are 3.2 eV, 1.9 eV, and 0.5 eV for the conduction edge of bulk Si and SiO_2, Si_3N_4, and SiC, respectively. Electrons effective masses of SiO_2, Si_3N_4, and 3C−SiC are 0.86, 0.05–0.13 [19], and 0.24 ± 0.1, respectively. Following this line of argument, the results suggest that dots in an SiO2 matrix would have to be separated by no more than 1-2 nm of matrix for a reasonable overlap of the wave function and hence of conductivity, while they could be separated by more than 4nm of SiC [10].

SILICON NANOWIRE SOLAR CELLS

The semiconductor nanostructures are hence proposed to combine with the organic materials to provide not only a large interface area between organic and inorganic components for exciton dissociation but also fast electron transport in semiconductors. Therefore, many research groups combined organic materials with semiconductor nanostructures to overcome the drawbacks of the organic solar cells. Many inorganic nanowires had been experimented for this purpose, including CdTe, CdS, CdSe,ZnO, and TiO2 nanowires [20]. To overcome this Silicon nanowires (SiNWs) have attracted much attention for photovoltaic applications because of their unique optical and electrical properties [21] and have the potential to impact many different technologies either through improved material properties or by offering a new geometry not possible with bulk or thin film devices [22]. It is known that SiNW arrays demonstrate excellent antireflection (AR) properties due to their broadband optical absorption by multiple scattering incidents, and therefore can be used as solar cell absorbers for trapping light [21]. In practical applications, a bunch of SiNWs are more useful than a single SiNW. The usage of SiNW arrays can be categorized into two types. The first one is for anti-reflection purposes. SiNW arrays are used to replace the conventional anti-reflection coating layer. The second one is SiNW-array core-sheath p–n junction solar cells. In particular, the second type combines the advantages of the first type, that is, a large junction area, a high anti-reflection ability, and a high light-trapping effect [23].

SiNWs can be prepared by fabricated by various techniques, including chemical vapor deposition (CVD), physical vapor deposition (PVD), reactive ion etching (RIE) combined with lithography techniques [22], dry etching, laser ablation, and vapor–liquid–solid(VLS) [20]. Among these, the metal-assisted chemical etching (MACE) technique is more facile and more economical to fabricate SiNW arrays since it avoids high temperature or high vacuum [22].

Fabrication of nanowire based solar cells with nanowire arrays formed on the entire wafer by the wet etch process has obtained a very reasonably good energy conversion efficiency. However, with the formation of SiNW arrays by VLS process on the entire silicon wafer, the solar cell performance is very poor and has an energy conversion efficiency of only 0.1% [24] due to improper doping condition used.

The SiNW arrays have been demonstrated as an efficient antireflection film for silicon solar cell by proper growth of density and length of wires [24]. Fabrications of these wires on si substrate, nanowire length has linear behave by etching time. Zeng et al. confirmed this with their experiments.The SEM observation clearly revealed that the lengths of the produced SiNWs increased with the etching time, ranging from 340nm to 1700nm, which indicates that the length of the SiNWs can effectively be tailored by prolonging the etching time. Fig. 19(f) shows the SiNW length as a function of the etching time. An excellent linear behavior can be obtained.

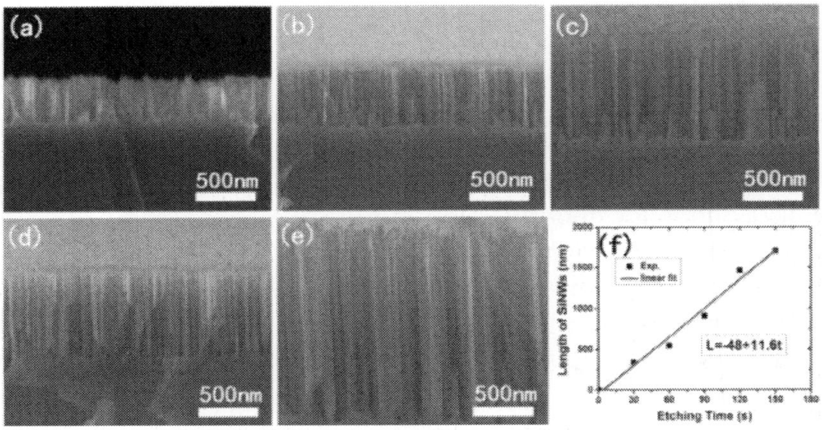

Figure 19: a-e) demonstrates the cross-sectional SEM micrographs of SiNW arrays with different length fabricated at 0.2 M H2O2 and room temperature for 30s, 60s, 90s, 120s, and 150s, respectively. (f) Shows the SiNW length as a function of the etching time [21].

Also the thickness of SiNWs has impressive on solar cell spectrum absorption and the optimum length of nanowires for 350 to 620 nm. Huang et al. in 2009 was confirmed on that by combination of the SiNWs and P3HT: PCBM blend is an attractive route to obtain high J_{sc} and efficiencies by improving the optical absorption, dissociation of excitons, and the electron transport.The P3HT:PCBM film exhibits little absorption beyond 650nm.The SiNWs/P3HT:PCBM film has improved light harvesting from 650 to 1100 nm because the cut off wavelength of Si is about 1100nm (Figure 20(Left)) [20].

Zeng et al. also shows that the as-grown SiNW arrays exhibit not only a large suppression of the reflectance over the entire light wavelength range, but also a very different reflection behavior from polished Si. The reflectance decreases with the wavelength increasing. As shown in figure 20(b), the reflectance is smaller than 18%, 9%, 5%, 2% and 1% at the nanowires length of 340nm, 542nm, 908nm, 1460nm and 1700nm, respectively. This can be attributed to the three important properties of SiNW arrays [21]:

1. The extremely high surface area of the SiNW arrays.
2. The suppression of the reflectance over a wide spectral bandwidth due to the subwavelength-structured (SWS) surface of the SiNW arrays.
3. A gradual change in the refractive index with depth due to a porosity gradient throughout the SiNW arrays which closely resembles a multi-antireflection layer coating.

It should also be mentioned that the interference peaks in the reflectance spectra of the SiNW arrays are related to the periodic nanostructure nature to some extent.

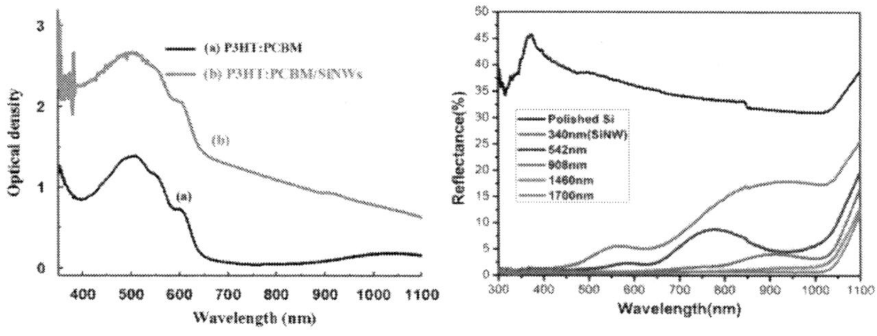

Figure 20: Left) UV–visible absorption of P3HT: PCBM blend on the ITO glass (a) with SiNWs and (b) without SiNWs. Right) Reflectance of SiNWs and polished wafer as a function of light wavelength, nanowires length of 340nm, 542nm, 908nm, 1460nm and 1700nm [20, 21].

Also the SiNW arrays grown on the silicon substrate with SiH4:N_2 gases exhibit good optical characteristic of antireflection with same result as mentioned [24]. On the other hand, it is also expected that the long SiNWs will have more light-trapping effect than the short SiNWs. The reflectance measurement does confirm this speculation because the reflectance decreases with the wire's length, as shown in Figure 20. However, this is not consistent with the devices' performance because the device with the shortest wire length has much better performance than the one with the longest length [23].

In addition, the reflectance decreases as the wire length increases. The reason can be easily explained by the enhanced light-trapping effect caused by the increasing wire length. However, this cannot explain the better performance of the device with a shorter wire length than that with a longer SiNWs [23].

Comparison of the photovoltaic performance for SiNW solar cells with SiNWs grown at different conditions. Figure21 show the device parameters, including the cell areas, and current density-to-voltage (J-V) curves under 1 sun AM 1.5 G illumination. The 0.37 μm-SiNW/PEDOT: PSS solar cell has the highest PCE of 8.40%, highest J_{sc} of 24.24 mA/cm^2, and highest V_{oc} of 0.532 V. When the SiNWs' length extends to 5.59 μm, PCE reduces from 8.40% to 3.76%, J_{sc} decays from 24.24mA cm 2 to 13.06 mA cm^2, V_{oc} decreases from 0.532 V to 0.435 V, and R_s increases from 2.95 O cm^2 to 4.25 O cm^2. Only fill factor (FF) stays in the range between 63% and 67% [23].

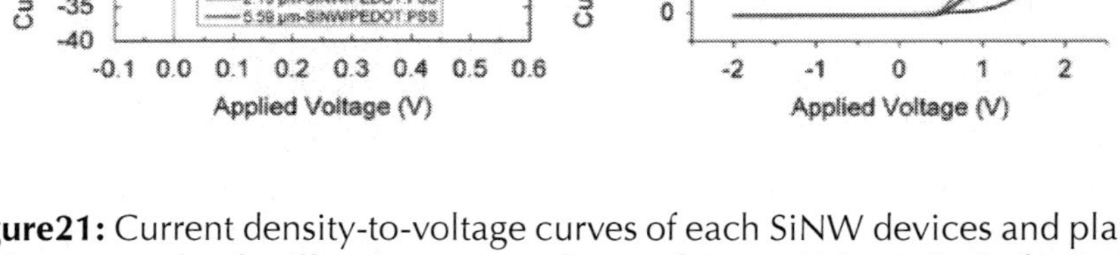

Figure21: Current density-to-voltage curves of each SiNW devices and planar Si devices under the illuminance condition of 1 sun AM 1.5 G. (Left) Curves around fourth quarter. (Right) Curves in the range of -2 V to 2 V [23].

However, the I–V characteristic measurements of this SiNW-based photovoltaic cell still indicate a much lower short circuit current, leading to much lower energy conversion efficiency in comparison with planar or textured Si solar cell. The much lower short circuit current is attributed to the lower density of the nanowire grown by the current VLS process, which could not provide efficient light trapping and causes large series resistance in the SiNW network [24].

At least the current SiNWs based solar cells have a much better performance than the ones reported in the literature, there is still much room to improve the performance by growing denser arrays of nanowires, which provide enough light trapping and reduce series resistance, and by growing nanowires with less amount of metal catalysts or other catalysts that do not cause recombination centers [24].

STRUCTURAL AND SUPERFICIAL OPTIMIZATION

There are a number of structural optimization methods that can improve the efficiency besides process optimization which is discussed in previous section. Such as surface texturing, anti-reflection coating, defect passivating by forming gas, use of concentrator system, etc. In this section the last two methods will be discussed more.

CONCENTRATOR PHOTOVOLTAICS

To moderate the price of multi junction solar cells and also to increase the efficiency, optical concentrator systems are proposed. The key elements of a photovoltaic concentrator are low-cost concentrating optics (a system of lenses or reflectors) to focus sunlight on a small area of solar cells, mounting, single or dual-axis tracking to improve performance of the system, and high-efficiency solar cells[14].

Theoretical maximum efficiencies of multi-junction solar cells without concentration and for concentration ratio of 500x have large difference. For example, the efficiency is 30.8% and 49.3% at one sun for one and three junction solar cell respectively, but it increases to 40.8% and 63.8% under concentration [14].

The first concentrator photovoltaic system was proposed in the mid 1970's by Sandia Labs. Despite the advantages of concentrating technologies, their application has been limited by the costs of focusing, tracking and cooling equipment. Optimization of a concentrator system is a complex problem: as all its components like solar cells, optics and tracking systems have to be specifically optimized, and all the interactions have to be regarded [14]. Natalya V. Yastrebova has reported several effective concentrator designs: Amonix is installing 250x concentrators using Fresnel lenses; Solar Systems, Spectrolab and Concentrating Technologies are installing reflective dishes; Sun Power is designing a high-concentration, thin (flat-plate-like) concentrator; the Ioffe and Fraunhofer Institutes are developing a 130x glass-Fresnel concentrator (Figure 22).

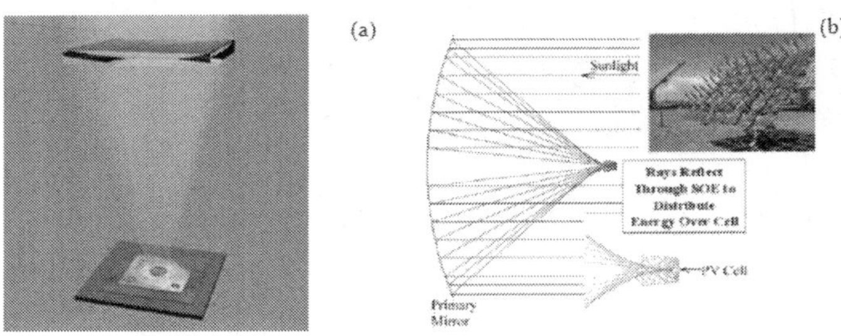

Figure 22: a) Concentrix concentrator using Fresnel lenses b) Spectrolab's reflective-optics concentrator module [14].

However, it should be taken into account that concentrators require direct sunlight and hence do not work with an overcast sky. Therefore, the concentrators are suitable for areas where cloud cover is low. The most appropriate time for operation is the middle of the day when the sunlight is strongest because at this time, the spectrum is least variable and hence spectral sensitivity is less significant [13].

Passivating the Interfacial Defects

One of the methods to enhance photoluminescence properties of nanostructured silicon based solar cells is passivating the non-radiative defects at the Si–barrier (SiO_2, Si_3N_4,...) interface by a forming gas (FG) post hydrogenation process performed on structures previously annealed in N_2 for different durations as proposed by Aliberti et al. A significant enhancement of the PL intensity has been observed for the samples with QD nominal size larger than 4nm, whereas a moderate enhancement is shown for the samples with QD size smaller than 4nm (Figure 23.a)[25].

Figure 23 b shows the PL spectra for two samples with a single layer of Si QDs in SiO_2 before and after FG hydrogenation. One sample has nominally 6 nm Si QDs and the other has 3.6 nm Si QDs. In addition to the intensity increase, which is more pronounced for the sample with larger QDs, no significant variation of the PL peak energy, or appreciable difference of the shape of the PL signal, can be observed in any of the samples [25].

The fact that the PL peak energy remains unmodified after the FG hydrogenation is an indication that, for sputtered single layer Si QDs in SiO_2, the PL cannot be attributed to defects. Therefore, the PL of these structures relies entirely on the quantum confinement

effect of Si QDs[25].

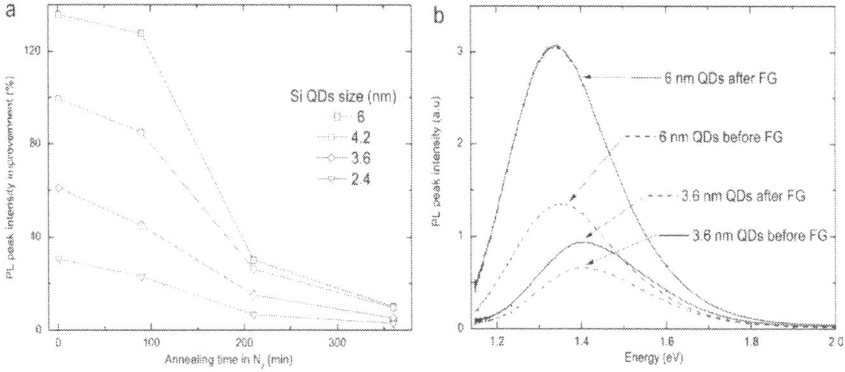

Figure 23: a) Relative improvement of the PL peak intensity for single layer Si QDs in SiO2 samples after FG post-hydrogenation versus annealing time in N2 (before FG) b) PL signal of two samples with different QDs nominal size (6 and 3.6 nm) before and after FG post-hydrogenation (samples have been annealed in N_2 at 1100°C for 5 min) [25].

CONCLUSIONS AND OUTLOOK

Third generation nanostructured silicon based solar cells offer significantly lower cost per Watt by applying multiple energy levels with abundant and nontoxic material that also benefits from thin film processes.

Therefore, optimization of these high efficient solar cells is a demand which should be satisfied by detailed researches. Then, in this chapter, various optimization methods have been taken into account.

The first is optimization in silicon quantum dot solar cells. We conclude that the control over QD size is possible for layer thicknesses less than about 7nm, within which Si migration to nucleating sites is dominated by a 2D rather than a 3D diffusion regime. For this layer thickness, the optimum size for Si QDs is 2-3 nm.

Moreover, to ensure favorable electronic transport, the optimum spacing should be satisfied. We conclude that in lower spacing, the possibility of percolation is enhanced; then the prominent regime will be charge transfer which is called migration instead of electron tunneling. It should be paid attention that different matrices produce different transport barriers between the Si QDs and the matrices, because tunneling strongly depends on the height of barrier. So, Si_3N_4 and SiC allow larger spacing for a given tunneling current in comparison with SiO_2 because they give lower barriers.

Although the electron tunneling conductivity is higher in SiC compared to Si_3N_4 and SiO_2 due to the lower barrier height (0.5eV) of SiC; the formation of Si-QD in SiNx is preferable, because the formation of 3-7 nm Si-QD in Si-rich SiC film requires higher thermal budget (1100°C) than Si-QD formation in a-Si/SiNx layer structure that requires lower annealing temperature (800-850°C).

Also, we conclude that Si QD fabrication by various vacuum deposition techniques is preferable because of the greater potential of integration into conventional devices.

The second is optimization in silicon nanowire solar cell which explains that in some aspects, Si NW is preferable in comparison with Si QD. The most important feature of SiNW is its crystallinity invariance under introduction of impurity atoms during the growth. In other words, SiNW is well-defined doped nanocrystal during synthesis. Moreover, it demonstrates ultra-high surface area ratio, low reflection, absorption of wideband light and tunable bandgap.

In addition, the absorption in Si NW is more than solid Si film. And in order to optimize SiNW, wire diameter, surface conditions, crystal quality and crystallographic orientation along the wire axis should be investigated. In practice, radial p-n junction NW cells tend to favor high doping levels to produce high cell efficiencies (up to 11%). Therefore, solar cells based on arrays of Si wires are a promising approach to reducing the cost of solar cell production.

However, SiNW has some disadvantages. For example, the probe light used in the optical measurement cannot be focused solely onto the nanowire.

In addition to above mentioned process optimization, structural optimization is discussed briefly. As a result, concentrator systems and forming gas can ensure the high efficiency nanostructured Silicon based solar cells.

This chapter brings an overview about various optimization methods which can be done over third generation nanostructured silicon based solar cells, however, there is a long way to achieve optimum values experimentally. So, more experimental researches in this area are required.

ACKNOWLEDGEMENTS

The authors want to thank INI (Iran Nanotechnology Initiative) Council and Student Scientific Association of University of Tabriz for their support. And, also they are very grateful to Dr. Rostami, Dr. Rasouli and Dr. Baghban who help them to do research in this field of photovoltaics.

REFERENCES

1. S Park, E Cho, D Song, G Conibeer, M. A. Green, 2009 Cells n -Type silicon quantum dots and p-type crystalline silicon heteroface solar cells. Solar Energy Materials & Solar 93684 690
2. G. Conibeer, 2007 Third-generation photovoltaics Material Today 10 11 44 50

3. P Löper, D Hiller, S Janz, M Hermle, S.W. Glunz, M Zacharias, 2011 23rd Workshop on Quantum Solar Energy Conversion 20-26 March 2011, Bad Hofgastein, Austria
4. A.K. Panchal, C.S. Solanki, 2011 Fabrication of silicon quantum dots in SiNx multilayer using hot-wire CVD. Journal of Crystal Growth 3112659 2663
5. F. Gourbilleau, C. Dufour, B. Rezgui, G. Brémond, 2009 Materials Science and Engineering B 15970 73
6. W. Z. Shen, 2011 Silicon Quantum Dots: Photoluminescence Controlling and Solar Cell Application. Seventh International Conference on Thin Film Physics and Applications
7. D Munteanu, J. L. Autran, 2011 Modeling of energy bands in ultra-thin layer quantum nanostructures for solar cell applications. Journal of Non-Crystalline Solids 357 1884 1887
8. C. Xu, Z. P. Li, W. Pan, W. Z. Shen, 2011 Tuning photoresponse through size distribution control of silicon quantum dots. Applied Surface Science 2578409 8412
9. C Jiang, M. A. Green, 2006 Silicon quantum dot superlattices: modeling of energy bands, densities of states, and mobilities for silicon tandem solar cell application. Journal of applied physics 99114902
10. E. Cho, M. A. Green, G. Conibeer, D. Song, Y. Cho, G. Scardera, S. Huang, S. Park, X. J. Hao, Y. Huang, L. V. Dao, 2007 Silicon Quantum Dots in a Dielectric Matrix for All-Silicon Tandem Solar Cells. Advances in OptoElectronics
11. H. Lan, Y. Ding, 2012 Ordering, positioning and uniformity of quantum dot arrays. Nano Today 7 94 123
12. D Di, I Perez-Wurfl, G Conibeer, M. A. Green, 2010 Formation and photoluminescence of Si quantum dots in SiO2/Si3N4 hybrid matrix for all-Si tandem solar cells. Solar Energy Materials & Solar Cells 94 2238 2243
13. G. Conibeer, M. Green, R. Corkish, Y. Cho, E. Cho, C. Jiang, T. Fangsuwannarak, E. Pink, Y. Huang, T. Puzzer, T. Trupke, B. Richards, A. Shalav, K. Lin, 2006 Silicon nanostructures for third generation photovoltaic solar cells. Thin Solid Films 511-512 654 662

14. N. V. Yastrebova, 2007 High-efficiency multi-junction solar cells: Current status and future potential. Centre for Research in Photonics
15. T. Nychyporuk, M. Lemiti, 2011 Silicon-Based Third Generation Photovoltaics. In: Kosyachenko L.A. (ed.) Solar Cells- Silicon Wafer-Based Technologies Rijeka: Intech 139 176 Available from http://www.intechopen.com/books/solar-cells-silicon-wafer-based-technologies accessed 2 November 2011
16. M. A. Green, G. Conibeer, 2007 Nanostructured Silicon-Based Tandem Solar Cells. PhD thesis. University of New South Wales, Sydney, Australia
17. D. Konig, J. Rudd, M. A. Green, G. Conibeer, 2009 Impact of interface on the effective band gap of Si quantum dots. Solar Energy Materials & Solar Cells 93 753 758
18. L. Pavesi, R. Turan, 2010 Silicon Nanocrystals: Fundamentals, Synthesis and Applications. Rijeka:WILEY-VCH Verlag GmbH & Co.
19. A. P. Goodey, S. M. Eichfeld, K. Lew, J. M. Redwing, T. E. Mallouk, 2007 Silicon Nanowire Array Photoelectrochemical Cells. Journal of American Chemical Society 12912344 12345
20. J. Huang, C. Hsiao, S. Syu, J. Chao, C. Lin, 2009 Well-aligned single-crystalline silicon nanowire hybrid solar cells on glass. Solar Energy Materials & Solar Cells 93 621 624
21. L. Zeng, X. Yu, Y. Han, D. Yang, 2011 Performance of silicon nanowire solar cells with a phosphorus diffused emitter. PhD thesis. Zhejiang University, Hangzhou, China
22. E. C. Garnett, P. Yang, 2008 Silicon Nanowire Radial p-n Junction Solar Cells. Journal of American Chemical Society 130 9224 9225
23. H. Syu, S. Shiu, C. Lin, 2012 Silicon nanowire/organic hybrid solar cell with efficiency of 8.40%. Solar Energy Materials & Solar Cells 98 267 272
24. C. Y. Kuo, C. Gau, B. T. Dai, 2011 Photovoltaic characteristics of silicon nanowire arrays synthesized by vapor-liquid-solid process. Solar Energy Materials & Solar Cells 95 154 157
25. P. Aliberti, S. K. Shrestha, R. Li, M. A. Green, G. J. Conibeer, 2011 Single layer of silicon quantum dots in silicon oxide matrix:

Investigation of forming gas hydrogenation on photoluminescence properties and study of the composition of silicon rich oxide layers. Journal of Crystal Growth 32784 88

Chapter 6

Going with the Wind: Temporal Characteristics of Potential Wind Curtailment in Ireland in 2020 and Opportunities for Demand Response

Eoghan McKenna[1], Philipp Grünewald[2], and Murray Thomson[1]

[1]Centre for Renewable Energy Systems Technology (CREST), School of Electronic, Electrical and Systems Engineering, Loughborough University, LE11 3TU, UK;

[2]Environmental Change Institute, Oxford University, Oxford, OX1 3QY, UK

ABSTRACT

The Republic of Ireland and Northern Ireland have ambitious targets for 40% of electricity to be supplied by renewables by 2020, with the majority expected to be supplied by wind power. There is, however, already a significant amount of wind power being turned down, or 'curtailed', and this is expected to grow as wind penetrations increase. A model-based approach is taken to estimate curtailment using high-resolution wind speed and demand data covering four years, with a particular focus on the temporal characteristics of curtailment and factors that affect it. The model is validated using actual wind output and curtailment data from 2011. The results for 2020 are consistent with previously published estimates, and indicate curtailment levels ranging from 5.6 to 8.5% depending on assumptions examined in this study. Curtailment is found to occur predominantly at night, and to exhibit stochastic variability related to wind output. To accommodate high penetrations of wind power, the findings highlight the value of flexible demand over relatively long time-periods. The model's output data have been made publicly available for free for further investigation.

INTRODUCTION

Integrating High Penetrations of Wind Power

With anticipated large penetrations of wind power, there is increasing concern about periods when there is too much wind power and not enough demand on the system. During such periods wind power can need to be turned down, also referred to as 'dispatched down'. Dispatch down can be required at the system level, for example, to maintain system stability, in which case it is referred to as 'curtailment', or it can be required locally, for example, to relieve local network congestion, in which case it is referred to as a 'constraint' on wind output.

Minimising or avoiding dispatch down improves the utilisation of wind energy, and as such can contribute towards decarbonisation goals. Methods to reduce dispatch down have been investigated as part of past studies concerned with integrating high penetrations of renewable energy onto the grid. Delucchi and Jacobson [1], in their evaluation of the feasibility of providing all energy for all purposes globally, suggest methods for dealing with high penetrations of intermittent renewables, including wind: creating diversity by geographically dispersing generators; using large-scale energy storage, either centralised, for example, hydro-electric facilities, or distributed, such as fleets of electric vehicles; hydrogen production during periods of excess generation; and demand response in the form of flexible heating and cooling demands.

Elliston *et al.* [2] found that it is technically feasible for 100% of electricity demand in the Australian New Energy Market to be met by renewable electricity, including 30% met by wind power. In their simulations, pumped hydro power was used to opportunistically charge during periods of excess renewable power. In a similar study, Budischak *et al.* [3] modelled combinations of wind, solar and electrochemical storage (batteries, and fuel cells) that could provide power for 90–99.9% of hours for the North American PJM grid. Looking at the pan-European grid, Rasmussen *et al.* [4] quantify the storage requirements for achieving a highly renewable powered European power system, and found this could be achieved with a combination of a relatively small, high-efficiency and short-term (6 h) storage, such as pumped hydro, and a larger, less efficient and long-term (seasonal) storage, such as hydrogen storage.

Wind Power Curtailment in Ireland

Although the studies above are concerned with methods for integrating wind in future scenarios, there are nonetheless present-day concerns with wind integration in the Republic of Ireland and Northern Ireland, where there is already a considerable amount of wind energy being dispatched down. In 2011, for example, 119 GWh of wind energy was dispatched down, equivalent to 2.2% of the total available wind output [5]. An estimated 80% of dispatch down in 2011 was due to curtailment, with the remaining 20% associated with constraints.

Although there are separate transmission system operators in the Republic of Ireland and Northern Ireland (Eirgrid and SONI, respectively), the two systems are interconnected and operate under a single electricity market. For the purposes of this paper, therefore, these will be referred to as a single 'All-island' power system [6].

The All-island system has ambitious targets for 40% of electricity to be generated from renewables by 2020 [7 , 8]. With one of the most abundant wind resources in Europe [9], it is expected that in 2020 wind power will account for almost 93% of renewable electricity. Thus 37% of electricity demand is expected to be met by 5 GW of wind capacity. For comparison, in 2012 2.1 GW of wind power had been installed [6].

The All-island system is relatively small, islanded and has low levels of interconnection with neighbouring power systems. Owing to the expected growth in wind power, dispatch down because of curtailment could increase considerably. Constraints on wind output are expected to be minimised in future because of on-going network reinforcements [10]. There is, therefore, an interest in estimating the level of curtailment of wind power in the future [11], as well as in investigating measures to reduce curtailment.

Estimates of Wind Curtailment in 2020

An early attempt at estimating curtailment came from the 'All Island Grid Study', which consisted of several technical reports commissioned by government to investigate generation portfolios for achieving 2020 renewables targets. As part of the study, Doherty estimated that curtailment would be negligible on the All-island system in 2020 for wind power penetrations below 5 GW, with curtailment rising to 14.4% of available wind energy with 9 GW of wind power [12]. Curtailment is estimated in the report using a system non-synchronous penetration (SNSP) limit of 66.6%. The SNSP is calculated as [13]

$$\text{SNSP} - \frac{\text{wind generation} + \text{HVDC imports}}{\text{system demand} + \text{HVDC exports}} \qquad (1)$$

Note there is a difference between (a) the instantaneous SNSP which will vary through time depending on the instantaneous levels of wind generation, demand and HVDC imports and exports, and (b) the SNSP 'limit', which refers to the threshold above which the instantaneous SNSP cannot be allowed in order to preserve system stability.

Where HVDC stands for high-voltage direct current power flows across the Ireland-GB interconnectors (see Section 2.4). As part of the same study, Meibom uses a mixed integer, stochastic, unit commitment and dispatch optimisation model to analyse the impact of increased wind penetrations [14]. No curtailment of wind power was found up to installed wind capacities of 6 GW.

Touhy and O'Malley [15] extended the model used in the All Island Grid Study to include the use of pumped storage as a measure to reduce curtailment and increase capacity factors for conventional fossil-fuel generators. Wind capacities ranging between 3 and 15 GW were considered, and system security limits were modelled by assuming a minimum of eight large generators were required on the system at all times. The results indicated that curtailment was negligible below installed wind capacities of 7 GW, and that while storage reduced curtailment levels, it was not economically viable for the wind penetrations that were considered.

Denny *et al.* [16] also extended the All Island Grid Study in order to consider the impact of increased interconnectors on the integration of high penetrations of wind power on the All-island power system, including the effect on curtailment. Curtailment was found to be negligible with installed wind capacity of 6 GW, and a doubling of interconnection with GB from 1 to 2 GW was found to reduce curtailment from 0.15 to 0.12%.

All of the above studies therefore found curtailment not to be an issue for the penetrations of wind power that can be expected in 2020. They all, however, used pre-recession growth estimates for demand, and since they were published these levels of demand are no longer correct. For example, Tuohy and O'Malley used a peak demand of 9.6 GW, and annual demand of 54 TWh, while current projections are for 7.3 GW of peak demand and 40 TWh for annual demand [6]. For comparison, peak demand in 2012 was 6.5 GW. As curtailment is dependent on the demand as well as the wind output, there is a need to revisit curtailment estimates given realistic estimates of demand growth.

Given the revised, post-recession demand growth estimates, McGarrigle *et al.* [11] estimate wind curtailment on the All-island power system, with the aim of determining required penetration levels to meet 40% renewable energy targets. A mixed integer programming, unit commitment and economic dispatch model is used, which included estimations for offshore wind output, and a simplified model of the GB power system. System security limits were included in the model based on published Transmission Constraint Group rules. Their results show that a SNSP limit of 60% resulted in curtailment levels of 14%, reducing to 7% for a SNSP limit of 75%. Owing to the use of revised demand growth estimates, this study indicates that curtailment is likely to be considerable on the All-island system.

Demand Response to Aid the Integration of Wind Power

The studies described above have considered various measures to reduce curtailment including storage, interconnectors and the relaxation of system security limits. Demand response is an alternative measure which is considered here. Demand response refers to consumers time-shifting electricity consumption in response to a signal, usually in the form of a financial incentive [17].

Keane *et al.* [18] investigated the use of demand response as a resource for the flexible operation of the All-island power system with high penetrations of wind power. A demand model was developed that characterises demand response resources as units that could be integrated into existing unit commitment models. They conclude that demand response can improve the reliability and efficiency of the system by reducing the use of conventional fossil fuel plant, although the impact on curtailment is not considered. Their study illustrates the potential importance of demand response on the All-island system. The focus of the study was, however, on the use of demand response for reducing peak demand, while this paper focuses on the potential for demand response to reduce curtailment by increasing demand during periods of excess wind generation.

The aim of this paper therefore is to investigate the potential for demand response to reduce curtailment on the All-island system. The next section describes the model that was developed to characterise the temporal characteristics that can be expected of curtailment in 2020. The model also provides estimates of total annual curtailment which can be compared to previous work.

MODEL OVERVIEW

Fig. 1 shows the high-level architecture of the model, which can be seen to consist of two main components: a wind power model and a curtailment model. The methodology for the model is based on the operating requirements faced by the system operator (as described in Section 2.3). The model does not account for unit commitment and economic dispatch of generators as (a) the paper is not concerned with the precise plant mix providing synchronous generation, so long and sufficient capacity is available, and (b) wind generation has the lowest short term marginal running costs of all major generators on the system and it is, therefore assumed that wind will be dispatched as much as possible within the constraints of the curtailments rules. The model was implemented in Matlab.

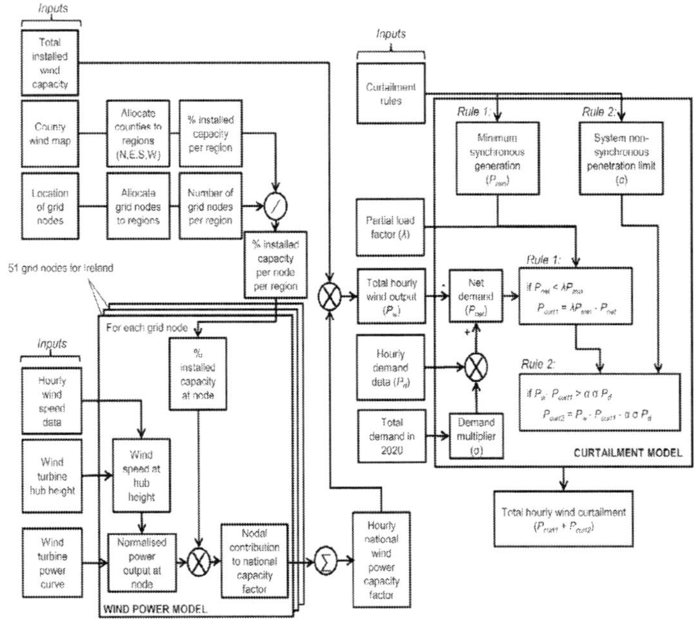

Figure 1: Model architecture for estimation of wind power and curtailment.

The wind power model takes an input that includes hourly wind speed data across a geographic grid, and produces aggregated hourly wind output for the whole island. The curtailment model takes as input the hourly wind output, hourly demand data and a set of curtailment rules to produce hourly estimates of curtailment. The temporal characteristics of the curtailment can then be analysed, and totals compared to previous work. The following sections describe the components of the model and data used in more detail.

Wind Power Model

To provide good temporal representation of curtailment, particular attention was given to developing an accurate estimation of the temporal variability of wind power on the All-island system. Similar to the approach taken in [19], high-resolution (half-hourly or hourly) historical wind speed data is used covering a period of several years. Owing to the use of multiple years of data, the model accounts for the distribution of wind speed over multiple years, and produces results that illustrate the possible range of future curtailment, rather than an estimate for a single year. This is a source of novelty compared with previous studies.

In addition to the requirement for several years of data, a good geographic spread was also needed to combine with the known geographic distribution of installed wind capacity in Ireland. Hourly wind speed data at 10 m for 2009 and 2010 was therefore sourced from the National Centers for Environmental Prediction (NCEP) Climate Forecast System Reanalysis (CSFR) product [20] covering the island of Ireland at 0.5° horizontal resolution, resulting in 51 grid nodes. For more recent data covering 2011 and 2012, the NCEP Climate Forecast System Version 2 (CSFV2) was used, which uses an identical model to CSFR, and can be considered a seamless extension to it.

Fig. 1 shows how the hourly gridded wind speed data is used to calculate estimates of hourly gridded wind output. Wind speeds at 10 m are scaled up to a hub height of 80 m using the power law method as

$$u_{80} = u_{10}\left(\frac{80\text{m}}{10\text{m}}\right)^{p} \qquad (2)$$

where u_x is the wind speed at height x. The p value depends on surface roughness and a value of 1/7 was used, as in [19 , 21]. The authors note that while the power law method for extrapolating wind speed is common, it has been shown to underestimate wind speeds at 80 m by an average of 1.3 m/s compared to a method that calculates vertical wind profiles based on a least-squares fit to twice-a-day wind profiles from sounding data [22]. Within the context of this paper, however, where the focus is on the temporal characteristics of the wind output, the use of the power law method is not inappropriate and, as will be demonstrated later, produces results that are validated against measured data.

Having calculated wind speeds at hub heights across the geographic area, a normalised wind output for each grid node ($P_{node,\ norm}$) is calculated based on the wind speed and a power curve based on the power curves of three commercially available large wind turbines. Note the model therefore assumes that all the wind turbines installed on the island are identical. This power curve is described in [23], and used in [24 – 26]. Turbine power output (P) and normalised wind output for each grid node ($P_{node,norm}$) are calculated as

$$P = \begin{cases} 0 \text{ when } u_{80} < 3 \\ 0.0015u_{80}^5 - 0.1967u_{80}^4 + 4.0058u_{80}^3 - 19.268u_{80}^2 \\ +43.607u_{80} - 50 \text{ when } 3 \le u_{80} \le 13 \\ 1000 \text{ when } 13 \le u_{80} \le 25 \\ 0 \text{ when } u_{80} > 25 \end{cases} \quad (3)$$

$$P_{node,norm} = \frac{P}{1000} \quad (4)$$

The normalised wind output for each grid node is converted to a nodal contribution to the (hourly) national wind capacity factor by multiplying it by the percentage of the total installed wind capacity that is allocated to the node (N_{node}). This is calculated as follows. The Irish County Wind Map provides installed capacities for each county [27]. The island is then divided into four regions (North, East, South and West), each county is assigned to one region, allowing percentages of total installed capacity to be calculated for each region (N_{region}). The number of grid nodes in each region is counted (n_{region}), allowing N_{node} to be calculated as

$$N_{node} = \frac{N_{region}}{n_{region}} \quad (5)$$

Note the model therefore assumes that within each region the wind turbines are evenly distributed between the region's grid nodes. Each nodes contribution to the national hourly capacity factor (C_{node}) is then calculated as

$$C_{node} = N_{node} \times P_{node,norm} \tag{6}$$

and the national hourly capacity factor ($C_{national}$) is calculated as the sum of the individual nodal contributions

$$C_{national} = \sum_{node} C_{node} \tag{7}$$

Finally, the national total hourly wind output (P_w) is calculated from the national capacity factor and total installed wind capacity ($N_{national}$) as

$$P_w = C_{national} \times N_{national} \tag{8}$$

For 2020, the regional distribution is assumed unchanged, and projected total installed wind capacity for the All-island system is taken to be 5061 MW [6]. The model does not account for offshore wind power, as only 0.2 GW is expected to be installed by 2020 [6].

Demand Data

Owing to the inherent time-correlation between wind and demand, concurrent demand data was required to estimate curtailment. Historic half-hourly demand data for 2009 through to 2012 was obtained from the Irish Single Electricity Market Operator (SEMO). This was then converted to hourly values and adjusted to account for day-light saving. Similar to the process adopted for wind, demand data is scaled up according to 2020 growth estimates [6]. Historic demand data is, therefore uniformly increased by 17% for 2020 estimates, as consistent with [11].

Note that the demand data is the total system demand and therefore includes demand that has to be met because of transmission and distribution losses. Losses are therefore accounted for implicitly in the model through the use of this data and are assumed to be unchanged compared to the historic data. Where demand is scaled up to 2020 levels, losses are therefore also assumed to scale proportionally.

Curtailment Model

The curtailment model is an implementation of curtailment rules taken from the 'Transmission Constraint Groups' and 'Operating Security Standards' documents [28], which detail the rules that govern which conventional generators need to be kept on-line for system security reasons. There are five system security limits described which necessitate curtailment [5]:

- system stability requirements,
- voltage control requirements,
- the SNSP limit,
- operating reserve requirements and
- morning load rise requirements.

The first two requirements (system stability and voltage requirements) have been implemented in the model using one rule. The Transmission Constraint Groups document gives the combinations of generators that can satisfy the two requirements, and the process adopted was to select the combination of generators that could satisfy all the requirements with lowest combined rated capacity. As in [11], the Transmission Constraint Groups are assumed unchanged out to 2020, because of uncertainty about which might be relaxed or changed. This results in three 'minimum synchronous generation' limits of 2503, 2258 and 2416 MW for week daytime, weekend daytime and night-time, respectively. These values, however, do not account for the fact that generators can be run at part load and so these limits are multiplied by a 'partial load factor' of 50%. In the model, wind power is curtailed if net demand falls below the minimum synchronous generation limits, as shown in Fig. 1 by the 'Rule 1' boxes.

The third requirement, SNSP limit, is implemented by its own curtailment rule. The model simulates three SNSP limits: the present limit of 50%, and two potential future limits of 60 and 75%, which are reported to become technically feasible in 2020 [13]. In the model, wind power is therefore further curtailed if the remaining wind power is greater than the demand multiplied by the SNSP limit as shown in the 'Rule 2' boxes in Fig. 1 .

The final two requirements are not separately accounted for in the model. It should be noted therefore that the model may under-estimate curtailment, particularly during morning periods when demand ramp up which may require additional reserve capacity.

In the model, the two curtailment rules are modelled as 'hard' limits, which cannot be violated. The alternative use of 'soft' limits, which can be violated but which might impose a price penalty, was applied by McGarrigle *et al.* [11] and resulted in negligible differences in curtailment levels.

Interconnection with the Gb Power System

Interconnectors can affect curtailment by allowing the All-island system to export to GB during periods of excess wind generation, as investigated by the authors [11 , 16]. Including interconnectors within the model is useful therefore as it allows the results to be compared with estimates from previously published studies.

The All-island system is interconnected with the GB power system via two interconnectors: the Moyle interconnector between Northern Ireland and Scotland, and the East-West interconnector between Ireland and Wales, both of which are rated at 500 MW in both directions. The number of interconnectors is expected to remain unchanged in 2020 [11], resulting in a potential 1 GW of export available to GB for the purposes of reducing curtailment.

It is not always possible, however, to predict whether export capacity will be available [6], as periods of excess wind generation on the All-island system are likely to coincide with periods of excess wind generation on the GB system. Furthermore, it should be noted that the Moyle interconnector has historically suffered from technical faults that limit its capacity [6]. The approach taken in the model was therefore to assume 500 MW of exports were constantly available to avoid curtailment.

MODEL OUTPUTS AND VALIDATION AGAINST HISTORIC DATA

Output of the Model over an Illustrative Week

Fig. 2 shows the output of the model over an illustrative week using data from January 2009. Demand and wind power are shown in Fig. 2 *a* and have been scaled up to 2020 levels. The week is characterised by generally high wind output apart from a calm during the middle of the week. Fig. 2 *b* shows the resulting net demand and the two curtailment variables which correspond to the model's two curtailment rules as described in Section 2.3. The first curtailment rule is associated with the minimum synchronous generation limits, which are shown by the grey line on the figure labelled as 'TCG' (Transmission Constraint Group). If net demand falls below this limit, then wind is curtailed, and the resulting total curtailment associated with this rule is illustrated in Fig. 2 *c*. Note that the use of interconnectors has not been included in these illustrations.

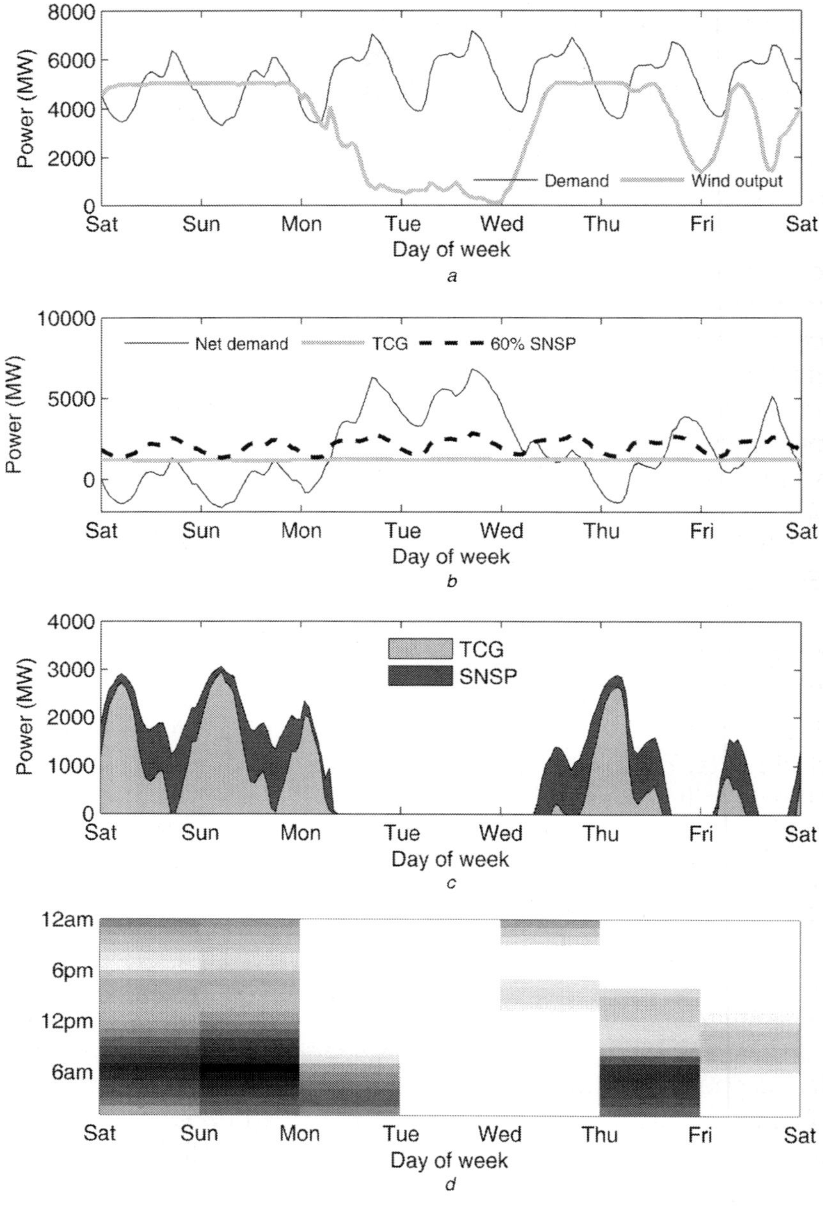

Figure 2: Operation of model over an illustrative week *a* Demand and wind output for one week in January *b* Net demand and curtailment variables *c* Total curtailment over one week *d* Alternative visualisation of curtailment.

The second curtailment rule is associated with the SNSP limit, which is taken to be 60% in this example. If the net demand after curtailment because of the first rule is below the demand multiplied by the SNSP limit, then wind is curtailed further. The resulting total curtailment because of this second rule is shown in Fig. 2 *c* as the black area.

Fig. 2 *d* provides an alternative visualisation of the total curtailment. Each day is represented as a column, with the start and end of the day corresponding to the bottom and top of the column, respectively. Each column comprises 24 'bins', one for each hour of the day. The colour of the bin represents the level of curtailment during that hour, with darker shades meaning more curtailment. This format highlights temporal patterns and variation of curtailment. In this example, four out of seven days were characterised by curtailment during the early morning hours and one period saw prolonged curtailment over two and a half days. We will return to the temporal analysis of curtailment in Section 3.3.

Validation of Wind Power Simulation

This section validates the model by comparing the model output with historic data. The wind power model will be discussed first, followed by the curtailment model in Section 3.3.

Fig. 3 plots the total hourly wind output from the model against empirical hourly data for wind generation for 2011. A black reference line is included to illustrate a perfect fit. Measured data was available for Republic of Ireland only and sourced from Eirgrid [29]. Installed wind capacity in the Republic of Ireland increased from 1374 MW in January 2011 to 1557 MW in December 2011 [30]. A limitation of the model, however, is that it requires a single value for installed wind capacity for the whole year. A time-weighted average of 1491 MW was therefore used. The model would appear to under-estimate wind output in the middle of the power range, and over-estimate wind output at the high end of the range.

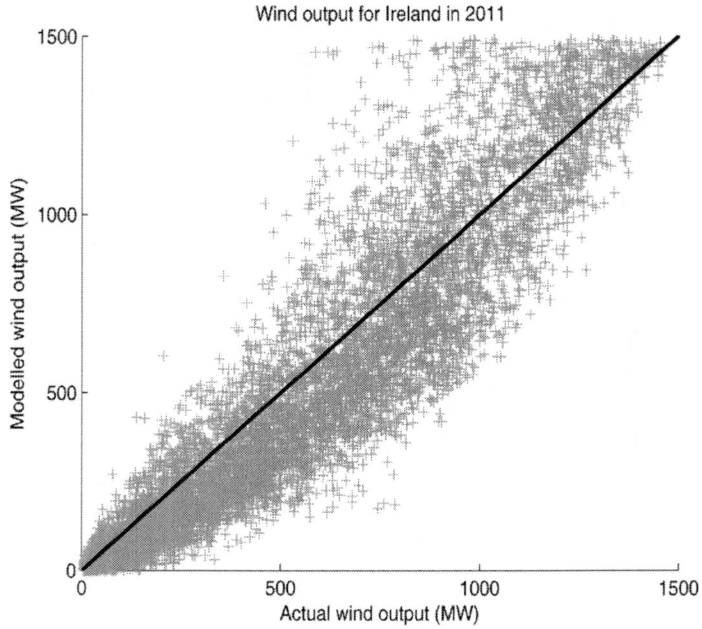

Figure 3: Modelled wind output using installed wind capacity of 1491 MW against actual measured wind output for Republic of Ireland in 2011.

To investigate this further, Fig. 4 shows the same data plotted as a probability density distribution. This confirms the overestimate for high outputs and the underestimate between 40 and 90%. It also illustrates that the model over-estimates wind output at the very low end of the power range. The capacity factor (C_{annual}) for the modelled data is 29.4%, which was calculated from the total modelled wind energy output for the year of the year (E_{tot}), and the total installed wind capacity ($N_{national}$) as

$$C_{national} = \frac{E_{tot}}{N_{national} \times 8760\text{h}} \quad (9)$$

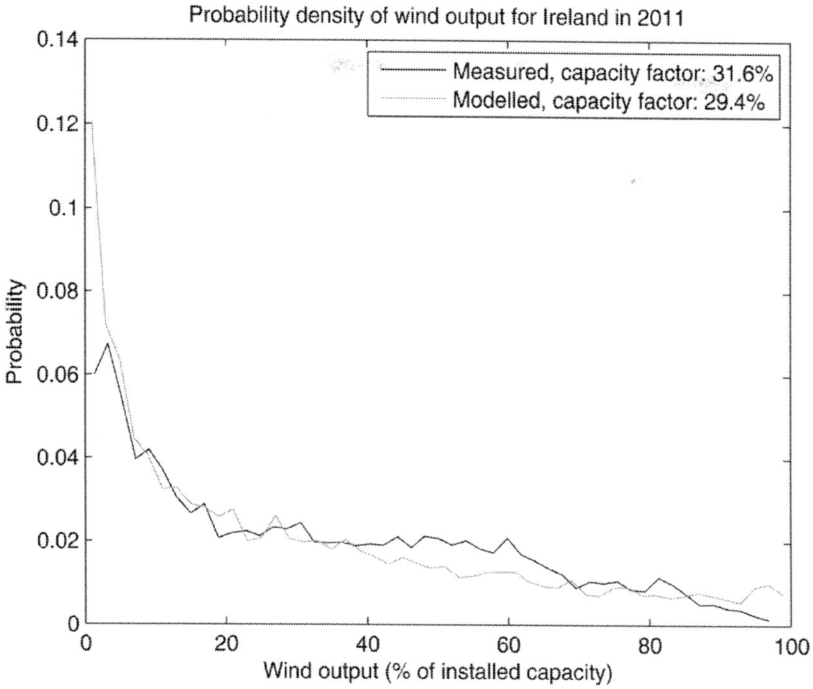

Figure 4: Probability density of modelled and measured wind output for Republic of Ireland in 2011.

This value can be compared to the capacity factor of the actual data (31.6%), which has been calculated as a time-weighted average of the reported monthly capacity factors for Republic of Ireland [30]. The authors note that the model underestimates the wind capacity factor by 7%, which could well be explained by the use of the power law method for wind speed extrapolation. The agreement between modelled and measured data is, however, reasonable for the purposes of this paper.

There is, however, a particular need for the model to accurately account for the temporal variability of wind output, so that the model's estimates of the timing of curtailment are realistic. Fig. 5 *a* shows an example of the modelled hourly wind output data compared with actual wind output data for the Republic of Ireland for May 2011. The model appears to capture the broad variations in wind output well, yet over shorter periods of time the model and empirical data can deviate.

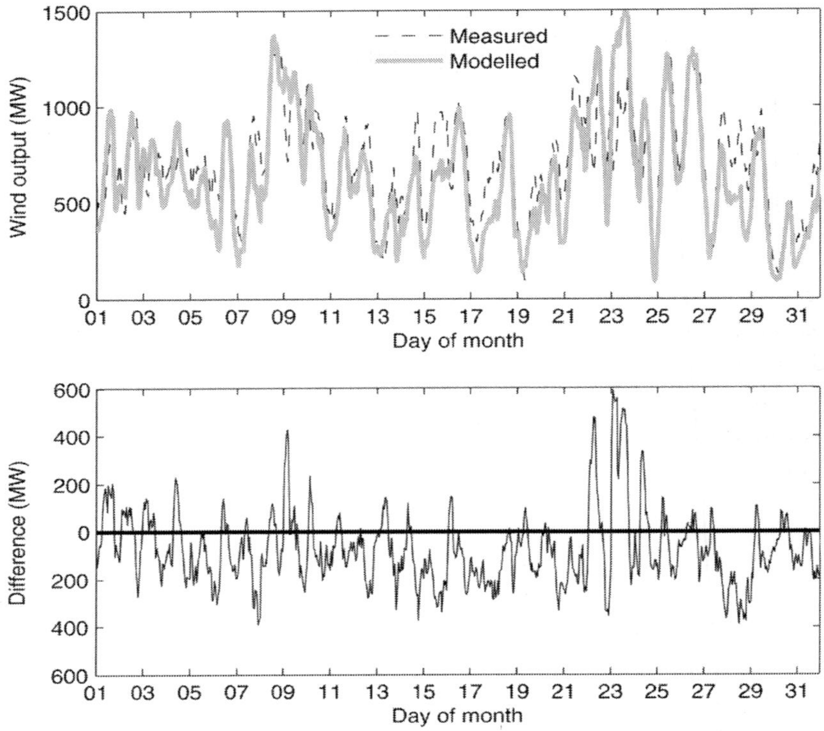

Figure 5: Measured and modelled wind output for All-island system for May 2011. Measured data from [29]. a Modelled and measured wind output for Iceland for May 2011 b Difference between modelled and measured wind output.

As this paper has a particular focus on accurately accounting for the timing of wind output and curtailment, it is important to explore the temporal accuracy of the wind model. Fig. 5 *b* shows the error between the modelled and measured data for the same month as Fig. 5*a*. The sum of the square of the error for the month shown in Fig. 5 was 1.05×10^{5} MW. For the wind model to be temporally accurate it is important to minimise the error between modelled and measured data. To check if the error between modelled and measured data is a minimum, the sum of the square of the error between modelled and measured data was calculated for the whole year. This calculation was performed using a range of temporal shifts to the modelled data, from 96 h lagging to 96 h leading, as

$$S_t = \sum_h \sqrt{\left(P_{model,h,t} - P_h\right)^2}$$

where s_t is the sum of the square of the error for time lag/lead t applied to the modelled data, $P_{model,h,t}$ is the modelled wind power output series for each hour h of the year with applied time lag/lead t, and P_h is the measured wind power output.

For the model to have relative temporal accuracy, the error s should be at a minimum at a temporal shift of zero hours, that is, when $s = s_0$, and for the error to increase as the modelled and measured data were increasingly mismatched in time (as $|t|$ increases). Fig. 6shows the relationship between the error and time lag/lead t. Errors have been normalised to 1, which occurs at the minima at 0 h shift, as

$$S_{norm,t} = \frac{S_t}{S_{t=0}} \qquad (10)$$

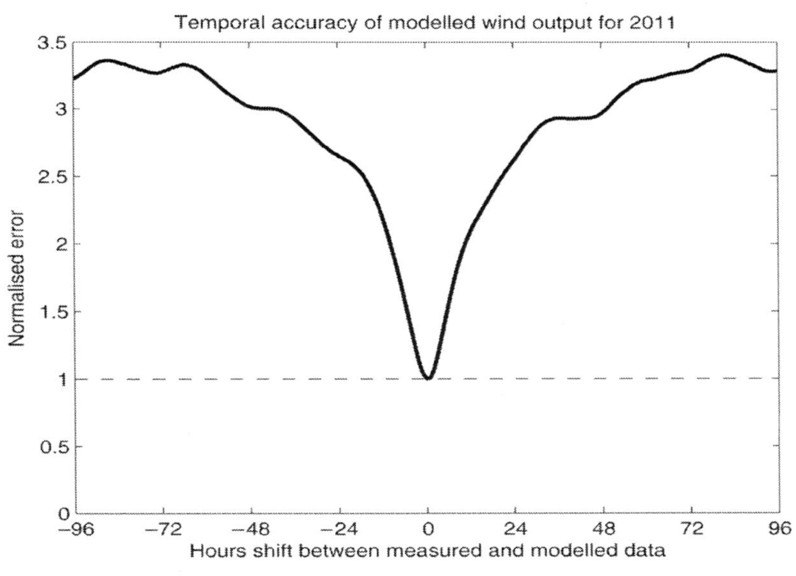

Figure 6: Temporal accuracy of modelled wind output compared to measured wind output for 2011

The scale of the error gives an indication of how well the timing of the modelled data matches the timing of the measured data. Local minima can be observed at 24 h intervals, and the error can be seen to level off after a period of about three days. The figure indicates that the model accurately accounts for the temporal variability of wind output on the All-island system.

Validation of Curtailment Model

Table 1 compares the annual curtailment results of the model against actual values for 2011. Annual wind output results are also shown. Actual values are taken from the 2011 Curtailment Report [5] and are All-island values. The value for total annual curtailment of 95 GWh is based on the report's estimate that 80% of the total of 119 GWh of dispatch-down was associated with curtailment. The modelled estimate for curtailment is 87 GWh, calculated using an assumed installed wind capacity of 1897 MW, which is based on a time-weighted average of installed wind capacity for the All-island system as reported in [30]. The model estimates for curtailment are 8 GWh less than the actual reported values, however, when taken as a percentage, the modelled curtailment results are in good agreement with the reported values (1.79% for modelled results compared with 1.82% for the measured results). This match should, however, not be taken as a measure of model accuracy. Results for other years may well deviate from reported values by greater amounts.

Table 1: Modelled wind output and curtailment data compared to actual values for 2011

	Modelled using 2011 data	**Actual 2011 values**
total wind generation, GWh	4880	5212
capacity factor, %	29.4	31.6
total wind curtailment, GWh	87	~95
percentage wind curtailed, %	1.79	1.82

Fig. 7 compares the sum of hourly curtailments for the model with actual values for 2011 for each hour of the day. The relatively modest amount of energy curtailment in 2011 is shown for comparison. Actual values (grey line) have been taken from the 2011 Curtailment Report [5] and are higher than the modelled values (black line), because they include constraints in addition to curtailment. The timing of model results appears to agree well with the actual values. The projected curtailment for 2020 in Fig. 7 will be discussed in Section 4.2.

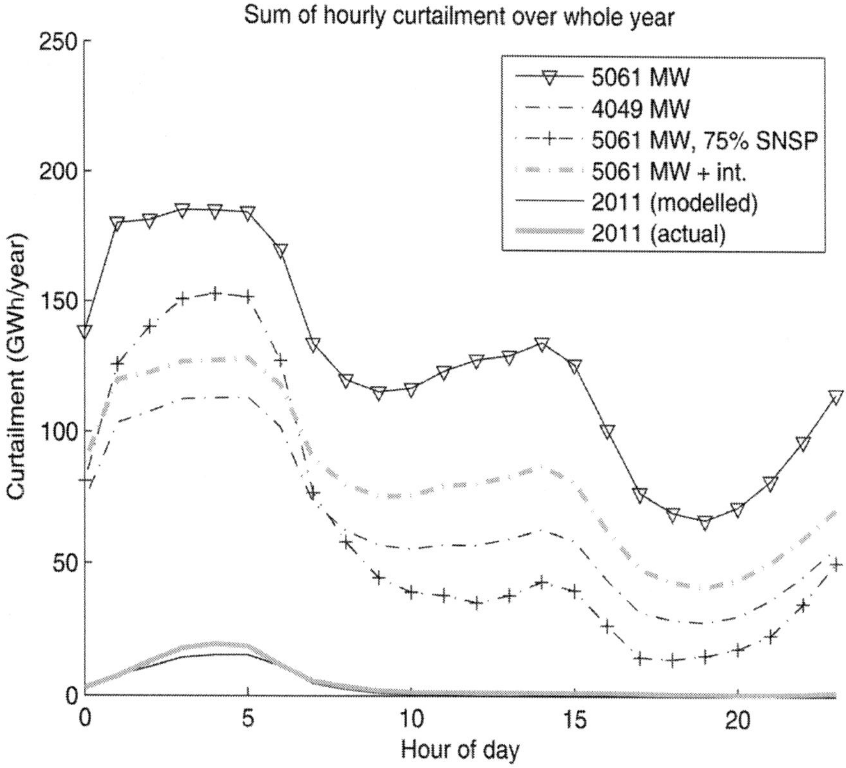

Figure 7: Distribution of hourly curtailment totals for different scenarios.

The previous sections have shown that the annual totals and timings of the results fit well against actual data, and that the model produces appropriate accuracy for the purposes of this paper. Having developed and validated this model, the next section looks forward to 2020 and characterises the timing of curtailment with a view to considering how to best make use of it with demand response.

RESULTS

Curtailment in 2020

Fig. 8 shows the results of the model simulation for 18 different scenarios, each with different assumptions for the year 2020. It illustrates the effect of installed wind capacity, SNSP limit, use of interconnection with GB, and yearly variation of wind output on curtailment levels. A 5 GW refers to the 5061 MW estimate for installed wind capacity in 2020 mentioned previously. The 4 and 6 GW cases are included to test the sensitivity of curtailment to higher or lower levels of wind capacity. Square data points in Fig. 8 represent mean values over a simulated range of four years, and the error bars the maximum and minimum annual values over this period. These results are also provided in table format in the Appendix.

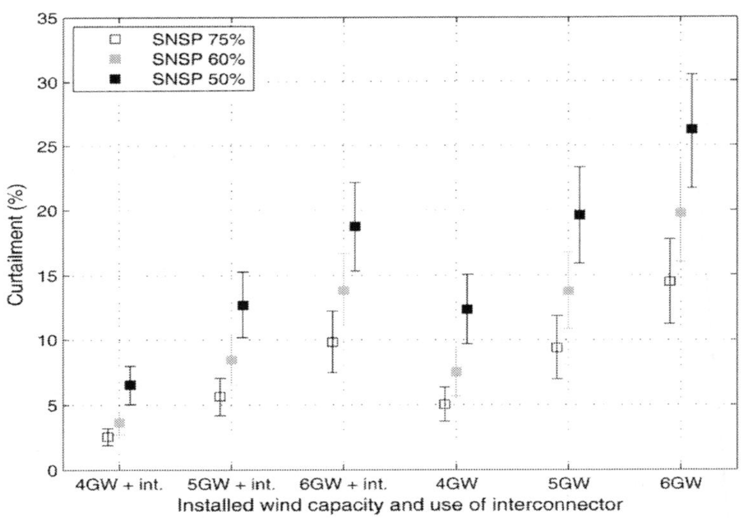

Figure 8: Effect of installed wind capacity, SNSP limit and interconnection with GB (+ int.) on curtailment levels on the All-island power system.

Curtailment can be seen to increase with increasing installed wind capacity, as well as with decreasing SNSP limit. The use of the 500 MW interconnector considerably reduces curtailment, based on the assumptions mentioned earlier. A 5 GW of wind with SNSP of 75% and use of interconnector results in a mean annual curtailment of 5.6%, and a maximum annual curtailment of 7.0% for the four individual years of data considered here. McGarrigle *et al.* [11] estimated curtailment levels between 6.5 and 7.3% for a comparable scenario of 5193 MW, SNSP 75% and use of interconnection. These values can be seen to be broadly within the range of values shown here.

Overall the results of the present study confirm the findings of McGarrigle *et al.* that there is likely to be a considerable amount of curtailment on the All-island system in the near future – provided nothing is done to address the problem. This highlights the importance of finding practical methods for reducing curtailment, one of which (demand response) is discussed in later sections.

Furthermore, note that the levels of curtailment are such that none of the scenarios explored here meet the target of 37% of demand being supplied by wind (see Table 4 in the Appendix for these results). The following section therefore considers the scope for demand response to improve the utilisation of wind energy.

Temporal Characteristics of Curtailment

To investigate the feasibility of the suggestion that demand response could reduce curtailment, it is important to understand when it occurs. Of the 18 scenarios presented in Fig. 8 , the following focuses on four, all of which have demand and wind scaled up to 2020 levels:

- Scenario 1: 5061 MW wind capacity, 50% SNSP
- Scenario 2: 4049 MW wind capacity, 50% SNSP
- Scenario 3: 5061 MW wind capacity, 75% SNSP
- Scenario 4: 5061 MW wind capacity, with use of 500 MW interconnector

Fig. 7 , mentioned earlier, illustrates when curtailment occurs throughout the day for the four scenarios. The distribution throughout the day is broadly similar, with more relative curtailment during the day for the 5061 MW 50% SNSP scenario. The 5061 MW 75% SNSP scenario shows that increasing the SNSP has more of an impact in terms of reducing curtailment during the day and less of an impact during the night and early morning. This is because curtailment during the night is predominantly associated with minimum synchronous generation limits (curtailment rule 1 as described in Section 2.3) and not because of the SNSP limit. All scenarios are characterised by a peak of curtailment occurring during the night and early morning, a levelling off during the day, and a trough of curtailment during the evening. Indeed the pattern that emerges is similar to an inverted system demand profile. The use of the interconnector can be seen to reduce curtailment, but not affect the timing of when curtailment occurs.

Although the aggregated view of curtailment shown by Fig. 7 reveals that curtailment occurs predominantly at night, it does not show whether this is consistent from one night to the next, in other words, how regular it is. This is a critical consideration for demand response purposes, and providing a clearer picture of the regularity of curtailment is a novel contribution of this paper. Fig. 9 , therefore, provides new insight into the timing of curtailment by showing hourly curtailment over four years for scenario 3 (5061 MW and 75% SNSP). Each hour of each day is represented by a pixel, with the shading of the pixel representing level of curtailment: black for the highest level, and white for zero.

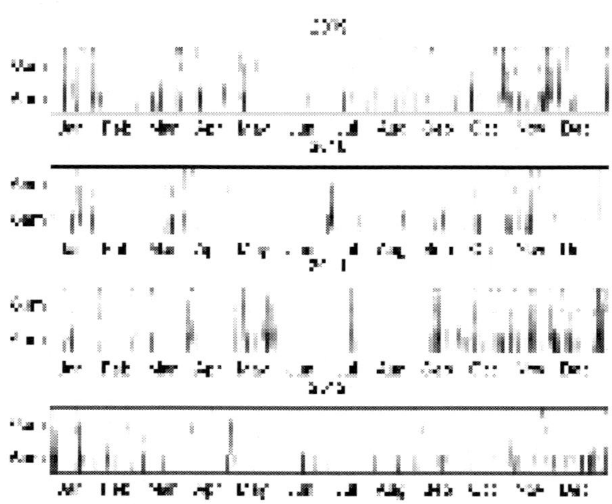

Figure 9: Curtailment events (dark areas) over a period of four years for 5061 MW wind capacity, 75% SNSP and no use of interconnector.

Four observations to curtailment can be drawn from the data for all of the scenarios. Firstly, curtailment occurs predominantly during the early hours of the morning as this is when demand is at its lowest. Secondly, curtailment tends to occur in 'clusters' that span several days. This is associated with the passing of large weather systems, which results in the variance of wind speed having a peak at a frequency of about four days [31]. Thirdly, curtailment has seasonal variability, with higher levels occurring during the winter months, when wind speeds tend to be higher and finally, there is considerable variability between years. Indeed, across all 18 scenarios considered here the average difference in total curtailment between maximum and minimum years is 57%. The data shown in Fig. 9 , along with other curtailment results from the model, have been made publicly available for further investigation [32].

DISCUSSION AND TOPICS OF FURTHER INVESTIGATION

Given the patterns of curtailment seen above, the obvious candidates for practical demand response applications are electric space and water heating. The higher level of curtailment during the winter months, for example, matches up well with the seasonal demand for heat. On a daily basis, demand for heat is during the day, whereas curtailment occurs predominantly at night. Thermal storage may therefore be required to shift heating demand. For space heating, this could be achieved using conventional electric storage heaters, or thermal heat pumps. Indeed it has been shown that demand associated with electrical heating in well-insulated dwellings with heat pumps could be shifted by up to 6 h without significant impact on occupant comfort levels [33]. Evidently, shifts of this scale would be useful for making use of curtailment for heating purposes.

Electric space heating is, however, not common in Ireland, and it is conceivable that it might be difficult to motivate consumers to invest in new electric heating systems. A recent survey, however, found that ~77% of households had electric immersion water heaters [34]. It is useful therefore to consider the volume of energy available because of curtailment compared to the volume of demand for water heating.

There are 1.66 million households in the Republic of Ireland [35], and 0.72 million in Northern Ireland [36], giving a total of 2.38

million. Assuming 77% have electric immersion heaters, makes 1.84 million hot water tanks. Assuming each tank holds 120 l of water, or ~120 kg, with a heat capacity for water of 4.2 J/gK, and a potential to be heated a further 20°C, gives 18.5 TJ or 5.14 GWh of energy that could theoretically be used to heat this amount of water once. For comparison, projecting forward to 2020, the 2011 data shown in Fig. 9 resulted in 1.54 TWh of energy curtailed over the whole year, or an average of 4.22 GWh per day, which is similar to the 5.14 GWh of available hot water storage. Technically, therefore, it would be possible to make use of some excess wind generation using existing assets such as electric immersion heaters, and this could mean that payback time of investment costs could be reduced.

Although the values for average curtailment per day and energy storage available in hot water tanks are similar, it should be noted that the average curtailment value masks the considerable variability of daily curtailment. Fig. 10 , for example, shows daily curtailment for the same scenario as above, and includes for comparison the 5.14 GWh hot water value. Although there are many days where curtailment could be used for hot water, there are evidently also many days where there is considerably more curtailment than could be feasibly used for this purpose. To reduce curtailment on these days, additional demand response measures would therefore be required.

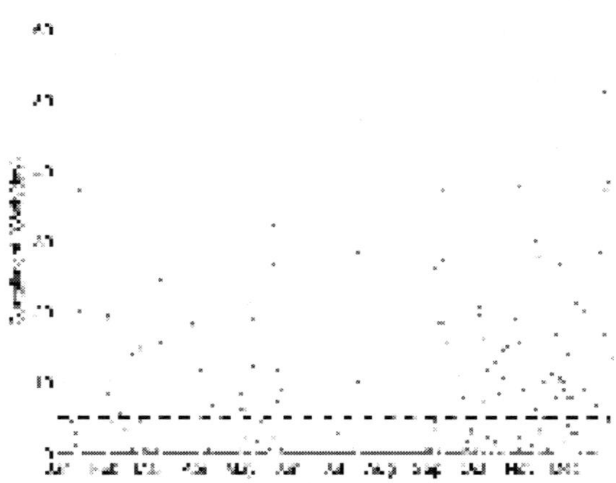

Figure 10: Daily curtailment totals for 5061 MW wind with 75% SNSP (2011 data).

The authors note that the predominant occurrence of curtailment at night could match well with future charging patterns of electric vehicles. The expected electrification of transport, by increasing demand at night, may well therefore provide an effective means for reducing curtailment, albeit one that would require considerable investment in terms of vehicles and charging points (unlike the use of electric immersion heaters which are already installed in 77% of dwellings).

A potential challenge highlighted by the results is that levels of curtailment vary considerably from year to year, which will introduce variability to the expected returns for consumers seeking to benefit from cheap electricity because of curtailment. A potential topic for future research therefore would be to quantify how much consumers might be expected to save on electricity bills given investment in technology to facilitate demand response to curtailment signals. It is also worth noting that while the conventional approach is to shelter the consumer from such forms of risk, it has been suggested that there is considerable scope to engage the consumer with the challenges of balancing the system [37].

More broadly the results of this paper highlight the value of 'dual-fuel' heating systems, for example, a gas boiler which can be used to provide space and water heating, as well as electric immersion heaters and portable electric radiators. Consumers with dual-fuel systems are able to shift back and forth to electricity depending on which fuel is cheapest at the time. The curtailment results shown previously indicate the availability of cheap electricity because of excess wind generation can vary stochastically, in which case there is value in being able to take advantage of cheap electricity when it is available, and switching to the alternate fuel during periods when wind output is low. This raises the important question for energy policy of whether we should commit to a more efficient but potentially less flexible all-electric future for heating, or focus more on developing dual-fuel systems, which are already relatively common in many people's homes.

CONCLUSIONS

This paper has presented a model for estimating curtailment on the All-island power system in 2020, using historic high resolution wind speed and demand data covering four years (2009–2012), and a set of curtailment rules. The model was validated using historic wind output and curtailment data for 2011.

For the four years of data considered here, estimates of mean curtailment levels in 2020 are in the range 5.6–8.5% for 5061 MW installed wind capacity and 500 MW of interconnection available for export to GB, across the expected range of SNSP limits (75–60%). In addition, this study has shown that none of the scenarios reach the target of meeting 37% of demand from wind and that demand response may be necessary to improve utilisation.

The majority of curtailment was found to occur during the night and in the early hours of the morning due to the low demand at these times. Increasing SNSP limit was found to reduce curtailment, but does so predominantly during the day, which still leaves the problem of considerable curtailment during the night and early hours of the morning.

Curtailment was also found to exhibit stochastic variability, for example, by occurring in 'clusters' lasting several days, occurring more during winter than summer, and by varying considerably from year to year. All of these are because of the fact that curtailment is dependent on the stochastic nature of the wind resource.

To make practical use of excess wind generation, electric water heating was found to be worthwhile given the existing high penetration of systems, and the planned roll out of smart meters which could provide the necessary communications infrastructure. There is the risk, however, that hot water tanks will not provide enough storage for clusters of curtailment over several days, in which case there is a value in solutions with larger storage capacities. 'Dual-fuel' heating systems, where electricity can be used for heat during periods of excess wind, and an alternate fuel used when electricity becomes expensive, could become valuable in such future systems.

ACKNOWLEDGMENTS

This work was supported by the Engineering and Physical Sciences Research Council, UK, within the HiDEF Supergen project (EP/G031681/1), the Transformation of the Top and Tail of Energy Networks project (EP/I031707/1), and the Realising Transition Pathways project (EP/K005316/1). The authors gratefully acknowledge data from the UK Meteorological Office, the National Centers for Environmental Prediction, Eirgrid Group, and the Irish Single Electricity Market Operator.

REFERENCES

1. DelucchiM.A.JacobsonM.Z.: ' Providing all global energy with wind, water, and solar power, Part II: Reliability, system and transmission costs, and policies', *Energy Policy*, 2011, **39**, pp. 1170–1190 (doi: 10.1016/j.enpol.2010.11.045) .
2. EllistonB.DiesendorfM.MacGillI.: ' Simulations of scenarios with 100% renewable electricity in the Australian National Electricity Market', *Energy Policy*, 2012, **45**, pp. 606–613 (doi: 10.1016/j.enpol.2012.03.011) .
3. BudischakC.SewellD.ThomsonH.MachL.VeronD.E.KemptonW.: ' Cost-minimized combinations of wind power, solar power and electrochemical storage, powering the grid up to 99.9% of the time', *J. Power Sources*, 2013, **225**, pp. 60–74 (doi: 10.1016/j.jpowsour.2012.09.054) .
4. RasmussenM.G.AndresenG.B.GreinerM.: ' Storage and balancing synergies in a fully or highly renewable pan-European power system', *Energy Policy*, 2012, **51**, pp. 642–651 (doi: 10.1016/j.enpol.2012.09.009) .
5. Eirgrid, SONI: '2011 Curtailment Report', 2012. Accessed 4 July 2013. Available from:http://www.eirgrid.com/media/2011%20Curtailment%20Report.pdf .
6. Eirgrid, SONI: 'All-Island Generation Capacity Statement 2013–2022', 2012. Accessed 16 September 2013. Available from:http://www.eirgrid.com/media/All-Island_GCS_2013-2022.pdf .
7. Department of Communications, Energy and Natural Resources: 'National Renewable Energy Action Plan', 2010. Accessed 11 September 2013. Available from: http://www.dcenr.gov.ie/NR/rdonlyres/03DBA6CF-AD04-4ED3-B443-B9F63DF7FC07/0/IrelandNREAPv11Oct2010.pdf .

8. Department of Enterprise, Trade and Investment: 'Energy: a strategic framework for Northern Ireland', 2010. Accessed 11 September 2013. Available from: http://www.detini.gov.uk/strategic_energy_framework__sef_2010_-3.pdf .
9. Risø National Laboratory: 'European Wind Atlas: European wind resources at 50 metres', 1989. Accessed 16 September 2013. Available from: http://www.windatlas.dk/Europe/landmap.html .
10. BurkeD.J.O'MalleyM.J.: ' Factors influencing wind energy curtailment', *IEEE Trans. Sustain. Energy*, 2011, **2**, pp. 185–193 (doi: 10.1109/TSTE.2011.2104981) .
11. Mc GarrigleE.V.DeaneJ.P.LeahyP.G.: ' How much wind energy will be curtailed on the 2020 Irish power system?', *Renew. Energy*, 2013, **55**, pp. 544–553 (doi: 10.1016/j.renene.2013.01.013) .
12. DohertyR.: ' All Island Grid Study. Workstream 2A: High level assessment of suitable generation portfolios for the All-island system in 2020', 2008. Accessed 16 September 2013. Available from: http://www.eirgrid.com/media/Workstream%202A.pdf .
13. Ecofys: 'All island TSO facilitation of renewables studies', 2010. Accessed 11 September 2013. Available from:http://www.ecofys.com/files/files/faciltiation_of_renwables_wp3_final_report.pdf .
14. MeibomP.: ' Final Report for All Island Grid Study'. Work-stream 2(b): Wind Variability Management Studies, 2007. Accessed 16 September 2013. Available from: http://www.eirgrid.com/media/Workstream%202B.pdf .
15. TuohyA.O'MalleyM.: ' Impact of pumped storage on power systems with increasing wind penetration'. Power & Energy Society General Meeting, PES '09. IEEE, Anonymous, 2009, pp. 1–8.
16. DennyE.TuohyA.MeibomP. et al.: ' The impact of increased interconnection on electricity systems with large penetrations of wind generation: a case study of Ireland and Great Britain', *Energy Policy*, 2010, **38**, pp. 6946–6954 (doi: 10.1016/j.enpol.2010.07.011) .
17. OwenG.WardJ.: ' Smart tariffs and household demand response for Great Britain', 2010. Accessed on 10 December 2012. Available from: http://www.sedc-coalition.eu/wp-content/uploads/2011/05/Sustainability-First-Smart-Tariffs-

and-Household-Demand-Response-for-Great-Britain-Final-March-20102.pdf .

18. KeaneA.TuohyA.MeibomP. et al.: ' Demand side resource operation on the Irish power system with high wind power penetration', *Energy Policy*, 2011, **39**, pp. 2925–2934 (doi: 10.1016/j.enpol.2011.02.071) .
19. GrünewaldP.CockerillT.ContestabileM.PearsonP.: ' The role of large scale storage in a GB low carbon energy future: issues and policy challenges', *Energy Policy*, 2011, **39**, pp. 4807–4815 (doi: 10.1016/j.enpol.2011.06.040) .
20. SahaS.MoorthiS.PanH. et al.: ' The NCEP climate forecast system reanalysis', *Bull. Am. Meteorol. Soc.*, 2010, **91**, pp. 1015–1057 (doi: 10.1175/2010BAMS3001.1) .
21. BestM.BrownA.HollisP. et al.: ' Small-scale wind energy'. A Met Office Technical Report, 2008. Accessed 18 September 2013. Available from: http://www.carbontrust.com/media/85174/small-scale-wind-energy-technical-report.pdf .
22. ArcherC.L.JacobsonM.Z.: ' Spatial and temporal distributions of U.S. winds and wind power at 80 m derived from measurements', *J. Geophys. Res., Atmos.*, 2003, **108**, p. 4289 (doi: 10.1029/2002JD002076) .
23. BartonJ.: ' A Probabilistic Method of Modelling Energy Storage in Electricity Systems with Intermittent Renewable Energy'. Accessed 18 September 2013 , 2007.
24. BartonJ.P.InfieldD.G.: ' A probabilistic method for calculating the usefulness of a store with finite energy capacity for smoothing electricity generation from wind and solar power', *J. Power Sources*, 2006, **162**, pp. 943–948 (doi: 10.1016/j.jpowsour.2005.07.006) .
25. BartonJ.P.InfieldD.G.: ' Energy storage and its use with wind power'. Power Engineering Society General Meeting, 2005. IEEE, Anonymous, 2005, vol. **2**, pp. 1934–1938.
26. BartonJ.P.InfieldD.G.: ' Energy storage and its use with intermittent renewable energy', *IEEE Trans. Energy Convers.*, 2004,**19**, pp. 441–448 (doi: 10.1109/TEC.2003.822305) .

27. IWEA: 'County Wind Map', 2013. Accessed 18 September 2013. Available from:http://www.iwea.com/index.cfm/page/_windenergy_onshore .
28. Eirgrid, SONI: 'Transmission Constraint Groups v1.2', 2012. Accessed 16 September 2013. Available from:http://www.eirgrid.com/aboutus/publications/ .
29. Eirgrid, 'System Performance Data: Wind Generation', 2013. Accessed 18 September 2013. Available from:http://www.eirgrid.com/operations/systemperformancedata/windgeneration/ .
30. Eirgrid: 'System Performance Data: All-Island Wind and Fuel Mix Report Summary 2011', 2011. Available from:http://www.eirgrid.com/operations/systemperformancedata/all-islandwindandfuelmixreport/ .
31. Vand.H.:'Powerspectrumofhorizontalwindspeedinthefrequency range from 0.0007 to 900 cycles per hour', *J. Meteorol.*, 1957, **14**, pp. 160–164 (doi: 10.1175/1520-0469(1957)014<0160:PSOHWS>2.0.CO;2) .
32. McKennaE.GrünewaldP.ThomsonM.: ' Wind Curtailment Estimates for Irish Power System in 2020 (Dataset)', 2014, Available from: https://www.dspace.lboro.ac.uk/2134/13302 .
33. HongJ.KellyN.J.RichardsonI.ThomsonM.: ' Assessing heat pumps as flexible load', *Proc. Inst. Mechanical Eng. A, J. Power Energy*, 2013, **227**, pp. 30–42. http://www.pia.sagepub.com/content/227/1/30.abstract (doi: 10.1177/0957650912454830) .
34. Commission for Energy Regulation: Electricity Smart Metering Customer Behaviour Trials (CBT) Findings Report. Accessed on 10 December 2012, 2011. Available from: http://www.cer.ie/en/information-centre-reports-and-publications.aspx?article=5dd4bce4-ebd8-475e-b78d-da24e4ff7339 .
35. Central Statistics Office: Statistics: Population, 2011. Accessed 29 September 2013. Available from:http://www.cso.ie/en/statistics/population/ .
36. NISRA: 'Household projections', 2011. Accessed 29 September 2013. Available from:http://www.nisra.gov.uk/demography/default.asp21.htm .

37. HigginsonS.ThomsonM.BhamraT.: ' For the times they are a-changin': the impact of shifting energy-use practices in time and space'. Local Environment, 2013, pp. 1–19.

Chapter 7

Thin Film Solar Cells Using Earth-Abundant Materials

Parag S. Vasekar[1] and Tara P. Dhakal[1]

[1]Center for Autonomous Solar Power, Binghamton University, Binghamton, NY, USA

INTRODUCTION

In a p-n junction photovoltaic (PV) cell, a photon of light produces an electron hole pair if the energy of the photon is at least equal to the band gap of the material constituting the p-n junction. The electrons and holes first diffuse toward the respective edge of the depletion region, and then drift across the junction due to the built-in potential and are collected at the electrodes. Thus, materials with long minority carrier life times and high carrier mobilities are desired for high efficiency. Because electrons have higher mobility than the holes, a p-type semiconductor is used as light absorber in a p-n junction solar cell.

The theoretical efficiency limit for an ideal homo-junction solar cell as calculated by Loferski [1] is 23%, and this maximum efficiency falls in the vicinity of the absorbers with band gap energy of 1.5 eV. A more justifiable theoretical efficiency limit that used atomic processes was put forward by Shockley and Queisser [2] later in the sixties. According to this theoretical efficiency limit, also known as the Shockley-Queisser limit, a maximum efficiency of 30% for a band gap of 1.1 eV is possible (assuming only radiative recombination) if exposed to the sunlight of global air mass 1.5.

The most recent data published in *Progress in Photovoltaics* [3] depict an efficiency of 25% for a crystalline silicon (Si) solar cell measured under the global AM 1.5 spectrum at 25°C. This value of efficiency approaches the theoretical value set by Shockley-Queisser limit. However, coupled with its low absorption cross section and high synthesis and processing cost, this first generation crystalline Si- solar cell doesn't show promise for a low cost, thin film PV device.

For thin film technologies, amorphous silicon-hydrogen alloy (a-Si) solar cells have exhibited efficiencies in the 10-12% range [4] and are fabricated with low cost technology. However, degradation of the performance over time is a major issue associated with the a-Si solar cells. Recently, metal induced crystallization of a-Si is used to make thin film polycrystalline Si-solar cell at relatively low temperature in an effort to reduce the processing cost [5], but fabrication of these devices is combined with high temperature chemical vapor deposition, which makes it incompatible with roll-to-roll processing. Other emerging second generation solar cells are thin film chalcogenides like CIGS ($CuInGaSe_2$) and CdTe, which show terrestrial cell efficiency of 20.3% and 16.5% respectively [6,7]. The scarcity, cost and toxicity associated with In, Ga, and Cd elements present in these cells limit their sustainability in the future.

Thus wide spread applications of solar cells will require dramatic decrease in cost through the use of non-toxic, inexpensive, and earth-abundant materials. The drawbacks in the present PV materials motivate us to look for alternatives. Due to the low absorption cross section, crystalline silicon requires thick layers, which will increase costs. Amorphous silicon (a-Si) has a higher absorption cross section and low processing cost compared to its crystalline counterpart, but it has stability problems. GaAs based solar cells have high efficiency, but the arsenic toxicity and the substrate cost limit their use. CIGS thin film solar cells show promise if the scarce and costly indium and gallium could be replaced by other elements. The quaternary compound semiconductor CZTS (Cu_2ZnSnS_4) [8] possesses promising characteristic optical properties; band-gap energy of about 1.5 eV and large absorption coefficient in the order of 10^4 cm^{-1}. The highly efficient CdTe solar cell is promising, but has obstacles such as cadmium toxicity, tellurium scarcity, and cost. While the results of research and commercialization of crystalline Si, GaAs, CIGS, CdTe, etc. cells are commendable, a search for alternative materials is indispensable and necessary to achieve low cost, light weight, and low toxicity arrays.

Table 1: Figure of Merits of CZTS, Zn_3P_2 and FeS_2 toward a PV cell compiled from the literature

Property	**CZTS**	**$Zn_{3P}2$**	**$Fe_{S}2$**
Band gap	1.5 eV [10]	1.5 eV [11]	0.95 eV [12]
Absorption Coeff.	□10^4 cm^{-1} [10]	□10^4 cm^{-1} [13]	3.3 x 10^5 cm^{-1} [12]
Electron diff. length	n/a+	4 to 10 μm [11]	0.13 to 1 μm [12]
Dark resistivity ( )	1 to 39 x10^3Ω. cm [14]	2.5 x 10^3 Ω.cm [15]	1.43 Ω.cm [16]
Carrier mobility (μ)	30 cm^2/Vs [10]	450 cm^2/Vs [17]	200 to 300 cm^2/Vs [12]

+Although the exact value of the electron diffusion length of CZTS is not available, the quantum efficiency (QE) spectra suggest a relatively lower carrier diffusion length for this material

In a recent study [9], a number of promising solar cell materials including CZTS, Zn_3P_2, and FeS_2etc. were identified as materials with low extraction cost. It is estimated that the cost of material extraction for CZTS, Zn_3P_2 and FeS_2 is around 0.005 cents/W, 0.0007 cents/W and 0.000002 cents/W respectively. More encouraging is the fact that the constituent materials of these absorbers are abundant in the earth's crust. The light absorption cross sections for CZTS, Zn_3P_2 and FeS_2

are all greater than 10^4 cm^{-1} (Table 1), thus making thin film devices practical. The band gaps are also in the optimum range for efficient photo-energy conversion. These three contenders will be discussed mainly in this review with some discussion on other earth-abundant candidates such as tin sulfide (SnS).

COPPER ZINC TIN SULFIDE

Introduction

Thin film solar cells based on Cu $(In,Ga)(S,Se)_2$ and CdTe have demonstrated significant improvement in the last few years, and they are also being transferred to production levels [18,19]. Out of these two technologies, CIGS based solar cells are the most efficient ones at the laboratory level and have demonstrated efficiencies in the range of 20% [6]. However, both CIGS and CdTe based thin film solar cells are hindered by potential environmental hazard issues [20] and scarcity issues associated with the constituent elements: mainly Te, In, Ga, and to some extent, Se [20, 21]. Recent research trends are moving towards finding alternatives based on earth-abundant and non-toxic elements. An alternative material $Cu(Zn,Sn)(S,Se)_2$ is being explored these days by the thin film photovoltaics community which contains earth-abundant materials like Zn and Sn. CZTS structure can be derived from $CuInS_2$ chalcopyrite structure by replacing one-half of the constituent indium atoms by zinc and other half by tin. The resulting bandgap varies in the range of 0.8 eV for a selenide structure to 1.5 eV for a sulfide structure [22]. Copper-Zinc-Tin Sulfide or Cu_2ZnSnS_4 (CZTS) has a nearly ideal direct bandgap (1.5 eV) to absorb the most of the visible solar spectrum as well as a high absorption coefficient (10^4 cm^{-1}). CZTS film contains neither rare metals nor toxic materials, and combined with the cadmium-free buffer layer, we can expect solar cells with complete non-toxicity.

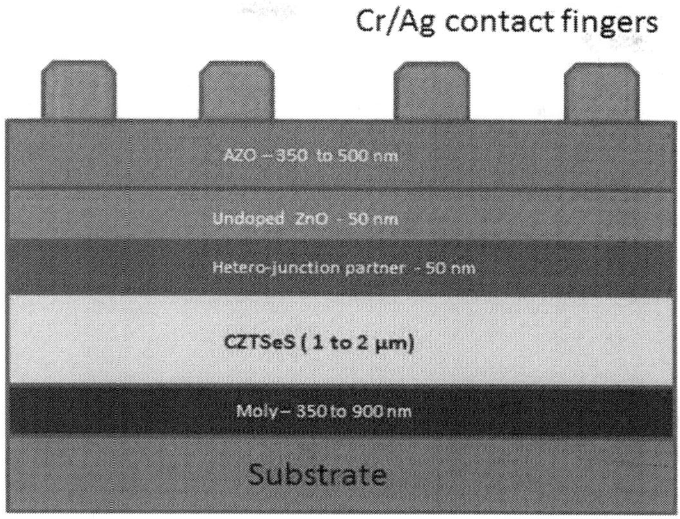

Figure 1: Schematic of the complete CZTSeS solar cell.

Several preparation methods have been reported in literature for preparation of CZTS solar cells [23]. These are physical vapor methods such as sulfurization of e-beam evaporated metallic layers and sputtered layers [24, 25], RF magnetron sputtering [26], co -evaporation [27], hybrid sputtering [28]. Chemical methods involve sulfurization of electrochemically deposited metallic precursors [29], photochemical deposition [30], sol-gel sulfurization methods [31], sol-gel spin-coated deposition [32] and spray pyrolysis [33, 34]. In addition, there are syntheses based on solution methods [35, 36].

Ever since Nakazawa reported the photovoltaic effect in a heterodiode of CZTS in 1988 [37], research in CZTS field has come a long way. The current highest efficiency for solution-processed CZTSeS based solar cells is 10.1% [38]. The efficiency for pure CZTS based solar cells that use the vacuum based approach is reported to be 8.4% [39]. The Schematic of a CZTSeS solar cell is as shown inFig.1.

Crystal Structure

CIGSeS based films are mostly crystallized with a chalcopyrite structure [40], while CZTSeS based films exhibit either a Kesterite type of crystal structure or Stannite structure [41]. Because XRD peaks for Kesterite structure are similar to Stannite structure, it is difficult to distinguish between them unless other measurement techniques, such as neutron diffraction or Raman spectroscopy, are also employed. All three crystal structures are shown in Fig. 2 in order to understand the similarity and difference between them. It is also quite likely that both Kesterite and Stannite structures co-exist simultaneously because there is not much of energy difference for these structures to achieve/attain a stable state [41].

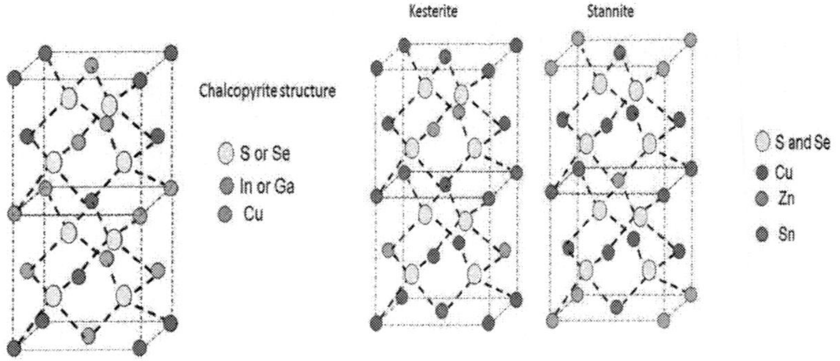

Figure 2: Chalcopyrite, Kesterite, and Stannite crystal structures.

Electronic Properties

The band-gap is the most important property of a photovoltaic material and near-optimum band-gap value around 1.5 eV is supposed to be ideal for effective photon absorption and photo-generation. CZTSe phase has a band-gap of 0.8 eV, while CZTS phase has a band-gap value of 1.5 eV [22]. The band-gap can be engineered between these two values by adding sulfur to CZTSe phase. The doping behavior in Kesterite is also controlled by intrinsic doping similar to their chalcopyrite counterparts. The p-type behavior of CZTSeS phase is controlled by Cu_{Zn} or Cu_{Sn} antisite donor defects, or this may also result due to direct creation of copper vacancy and compensation for Zn_{Cu} antisite and the S vacancy [41].

Phase Stability

Similar to CIGSeS thin films, CZTSeS phase also has a narrow region of phase stability in which the device quality phase can be synthesized without any adverse effect of secondary phases. In fact, the phase stability region is even narrower as compared to chalcopyrites, thus making single phase device quality CZTSeS synthesis more difficult [42]. Solid state CZTS can also be synthesized through reactions between ZnS, Cu_2S, and SnS_2. As can be seen in the phase diagram in Fig. 3, there is very narrow region where CZTS is stable and secondary phases are very easy to form. Copper-poor and zinc-rich composition has been found to be ideal so far to obtain efficient devices from CZTS films [8].

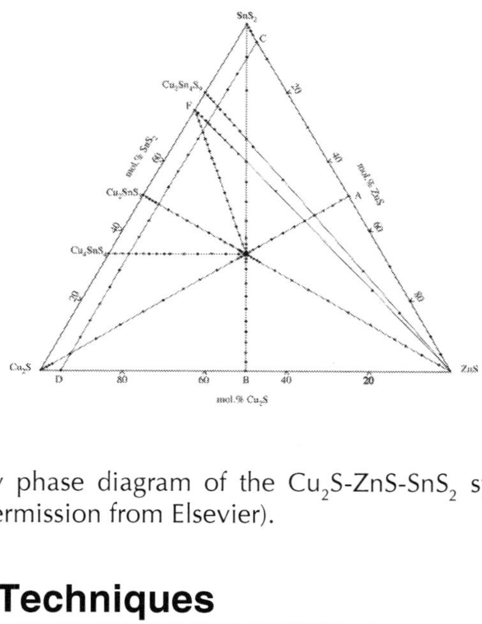

Figure 3: Ternary phase diagram of the Cu_2S-ZnS-SnS_2 system. (Reprinted from [42] with permission from Elsevier).

Synthesis Techniques

Vacuum Based Approaches (Sputtering/ Evaporation)

The synthesis techniques are both vacuum-based processes and non-vacuum based processes. Dr. Katagiri's group has pioneered vacuum-based approaches [8, 22 ,24 ,43]. Their first report on E-B evaporated precursors followed by sulfurization yielded an efficiency of 0.66% [24]. With modification of fabrication process, the conversion efficiency was gradually increased in their successive reports. It was observed that longer time and temperature cycles were detrimental for the performance of the CZTS films: hence, by adjusting the time and temperature cycles, Katagiri et al reported an efficiency of 2.62% in their next report [14]. After overcoming the problem of residual gases in the sulfurization chamber, the efficiency number rose to 5.45% [44].

This group has also contributed towards sputtered CZTS films followed by sulfurization [43] and reported an efficiency of 6.7%. They observed that soaking the CZTS layer on the Mo coated sodalime glass in deionized water leads to significant improvement in the device performance. Recently, they also reported 6.48% efficiency using CZTS compound target and a simple single sputtering step [45].

In 2010, IBM group reported an efficiency of 6.8% by using co-evaporated film of Cu, Zn, Sn and S sources followed by reactive annealing [46]. The best efficiency so far for pure sulfide phase CZTS is 8.4% [39], as noted in the IBM group report for evaporated CZTS films. Also, they recently reported an efficiency for pure selenide phase using co-evaporation is 8.9% [47]. An NREL group reports the best efficiency for pure selenide phase CZTSe as 9.15% [48]. The highest reported efficiency using vacuum based techniques is 9.3% [49], which is for mixed CZTSeS phase using co-sputtering of compound targets followed by annealing at a higher temperature. This report is by an industrial research group (AQT Solar).

Some other notable recent reports on vacuum based approaches are as follows:

1. In the quest to obtain Cd-free devices, the researchers at Solar Frontier, Japan recently demonstrated an efficiency of 6.3% for In-based hetero-junction partners and 5.8% for Zn-based hetero-junction partners [50].
2. The same group in Japan reports > 8% efficiency on CZTS sub-module [51]. Another interesting and note-worthy contribution here is that the absorber thickness is reported to be just 600 nm.
3. Salomé et al demonstrate that incorporation of H_2 is beneficial for preventing Zn loss during the sulfurization [52].

4. Shin et al mention that stacking of precursors has a role in getting single-phase final CZTS compound and reducing the secondary phases [53].
5. Chalapathy and co-workers demonstrate an efficiency of 4.59% using ZnSn

Electro-deposition is of particular interest for the thin film photovoltaics community due to the advantages it offers: low cost, environment friendly, large area deposition, room temperature growth, and less or almost no wastage of materials. CIGS technology based on electro-deposition has already been commercialized [55]. There are numerous reports on electro-deposition of CZTS as well [29, 56, 57]. Most of the reported efficiencies are in the range of 3% to 4% for electro-deposited CZTS films [57-59].

Some other notable recent reports on electro-deposited CZTS films are as follows:

1. Li et al recently demonstrated 1.7% efficiency in CZTSe films prepared by co-electro-deposition [60].
2. Mali et al used a successive ionic layer adsorption and reaction (SILAR) method for CZTS films. In this method, a substrate is immersed into separately placed cations and anions. This particular electro-chemical method is used for making uniform and large area thin films [61].

Spray Pyrolysis

Spray pyrolysis is a versatile as well as a low-cost technique that has been used to deposit semiconductor films. In this process, a thin film is deposited by spraying a solution on a hot surface. The constituents react to form a chemical compound. The chemical reactants are selected such that the products other than the desired compound are volatile at the temperature of deposition [62]. There have been early reports of sprayed CZTS films [33].

Some other notable recent reports on sprayed CZTS films are as follows:

1. Prabhakar et al prepared CZTS films using ultrasonic spray pyrolysis and observed that sodium plays an important role in CZTS films as well [63]. This observation similar to CIGS films

and the role of sodium in CIGS films is well known. The authors observed that sodium enhanced (112) orientation and grain size of the CZTS films also increased the hole carrier concentration.

2. In another report, authors prepared CZTS films using two different tin precursors: stannous as well as stannic chloride [64]. They observed that crystallinity and grain size as well as carrier concentration and mobility values were better for films prepared using stannic chloride.

Solution Based Methods

Direct liquid deposition or 'ink' based approaches are attractive due to their compatibility with high volume manufacturing techniques such as printing. It is a low-cost scalable route for the thin film solar technology. One of the popular approaches is to prepare precursors using sol-gel method, then sulfurize it, and deposit on substrate using spin-coating [65]. The micro-particles of CZTS have also been prepared using ball milling, sintering type of processes and later screen printed on flexible substrates [66]. There are various other reports using colloidal nanoparticles [67], Photochemical deposition [30] etc.

The researchers at IBM in their recent publication have reported the highest record efficiency by any process so far for CZTSeS as 10.1% [38]. The J-V curve of this record sample and cross-sectin of the CZTSeS film is as shown in Fig. 4. Here metal chalcogenides were dissolved in hydrazine solution and later spin-coated on glass-substrates followed by a heat-treatment at 540°C.

Some other notable recent reports on solution/ink based CZTS films are as follows:

1. Using band-gap engineering with Germanium, researchers at Purdue University have achieved an efficiency of 8.4%. They have utilized nanoparticles of CZTSeS on Moly-coated glass substrates [68].
2. The researchers at DuPont have demonstrated an efficiency of 8.5% [69]. They use binary and ternary chalcogenide nanocrystals produced by simple colloidal syntheses.

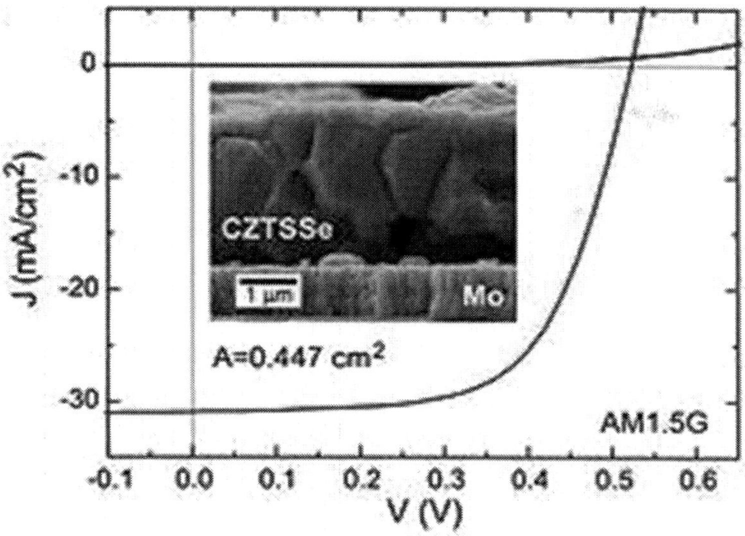

Figure 4: J–V characteristics of the record 10.1% sample at AM1.5G simulated illumination along-with CZTSSe film cross-section (Reprinted from [38] with permission from John Wiley and Sons).

Current Status

The laboratory level efficiencies are more than 10% [38] and there are some recent reports of sub-module efficiencies of more than 8% [51]. However, the technology still needs to go a long way before it can be commercialized. The complex phase diagram and material properties impose most of the challenges. Also, the narrow stoichiometry window for a stable CZTS phase requires a robust control on composition. The processing window is even narrower compared to their CIGS counterparts. The volatile nature of some components, such as tin sulfide and zinc, impedes further challenges for traditional vacuum-based processing approaches which require high sulfurization/selenization temperatures after precursor deposition. The secondary phases formed are difficult to detect with conventional techniques such as XRD due to the overlapping peaks with pure CZTS phase. The interface properties with hetero-junction partners need to be studied well. However, the initial results in last few years look quite promising.

ZINC PHOSPHIDE (ZN_3P_2)

Introduction

With rise in the prices and non-abundance of the materials such as indium and gallium current research trends in thin film solar cells have been moving toward development of earth-abundant solar cell materials that can be synthesized using low-cost processes. Also, zinc-based hetero-junction partners are preferred over toxic cadmium based compounds such as Cadmium sulfide [70]. Zinc phosphide (Zn_3P_2) is an important optoelectronic material, which also has applications in lithium ion batteries [71]. It is an important semiconductor from the II-V group, and is used for optoelectronic applications [13,72-74]. Zinc phosphide exhibits favorable optoelectronic properties, such as direct bandgap of 1.5 eV, which corresponds to the optimum solar energy conversion range [74-77]. Zinc phosphide has a large optical absorption coefficient of $>10^4$ cm^{-1}, hence it can be positively used as a p-type absorber [78]. Also, due to its long minority diffusion length of ~10 μm, high current collection efficiency can be yielded. Zinc, as well as phosphorous, is abundant in the earth's crust. This makes their cost-effective development quite feasible when it comes to large scale production [73,79]. Zinc phosphide is a tetragonal p-type low cost material with lattice constants of a = b =8.097 Å and c=11.45 Å [17] and has all the right characteristics for photo conversion. Zn_3P_2 is synthesized most of times with a p-type conductivity [80]

Early Efforts

Thin film solar cell devices using Zn_3P_2 have been fabricated using Schottky contacts, p-n semiconductor hetero-junctions [81] or liquid contacts [82]. Zinc phosphide was explored extensively in the early eighties and nineties [11, 78, 80, 83 ,84]. With a Schottky diode, an efficiency as high as 6% was demonstrated [78]. Zinc phosphide homo-junctions have been difficult to make. However, even a zinc phosphide hetero-junction solar cell has not been successful with an efficiency range around 2% [11, 85, 86].

The researchers at the Institute of Energy Conversion were leaders in the early eighties in exploring Zn_3P_2 [78,80,83,87,88]. They demonstrated the effect of Mg doping and also fabricated Schottky barrier solar cells [78,88]. They also successfully made hetero-junction solar cells using ZnO as a hetero-junction partner [11].

Another interesting study is by Misiewicz et al [89]. They made various metal contacts (Au, Ag, Sb, Al, Mg) with Zn_3P_2 and measured current-voltage characteristics. It was observed that Ag and Sb made the best ohmic contacts, while Au, Mg and Al exhibited rectifying properties. Mg was found to be the most useful for making Schottky barriers.

Synthesis Technqiues

Some of the synthesis techniques to grow Zn_3P_2 are summarized below:

Vapor Transport

Catalano et al [90] report synthesis of Zn_3P_2 by vapor transport using perforated capsule technique. Crystals were grown in silica tubes placed mid-way in a capsule immediately ahead of Zn_3P_2 charge. When heated at a high temperature (900°C), effusion of zinc and phospherous out of the inner growth capsule resulted in the condensation of Zn_3P_2 on the walls of outer tube's cooler portion. A similar vapor transport technique has been reported by Wang et al [91] for growing single crystal Zn_3P_2.

Evaporation

Deiss et al report synthesis of Zn_3P_2 thin film by direct evaporation of Zn_3P_2 on glass substrates in a vacuum chamber at 750°C [92]. The films deposited on a cold substrate were reported as amorphous, while the films deposited on a heated sunstrate (>300°C) were reported as crystalline. In a similar technique, Murali et al report that films grown with source-substrate distance < 3 cm were crystalline in nature [93].

Metal Organic Chemical Vapor Deposition (MOCVD)

Kakishita et al report a low-pressure MOCVD system with dimethylzinc (DMZ) and diluted phosphine (PH_3) as reactant gas [72]. Phosphine was cracked at 800°C and Zn_3P_2 films were grown on ZnSe single crystal substrates at a growth temperature of 290-410°C. Hermann et al report use of diethlzinc (DEZ) as a zinc precursor [15].

Electrodeposition

Soliman et al [94] report electrodepositon of Zn_3P_2 with SnO_2 coated glass as working electrode and the aqueous solution containing different concentrations of both $ZnCl_2$, $ZnSO_4$ and Na_2PO_3. The pH was adjusted at 2.5 by adding H_2SO_4. It was observed that $Zn_3P_2$was obtained, and the degree of crystallinity increased with higher Zn/P molar ratio. There is a recent report similar to that of Soliman et al, reporting electrodeposition of Zn_3P_2 [95].

Chemical Reflux Technique

Recently our group has successfully synthesized Zn_3P_2 using chemical reflux technique [96,97]. The zinc-coated glass substrate is placed on the top of a fritted glass, which is sorrounded by trioctylphosphine (TOP) liquid. The entire vessel is heated while vapors of TOP are refluxed back to the bottom of the glass vessel by a condenser connected at the top of the vessel. The glass vessel is heated to 350°C for two to four hours, and then slowly cooled to room temperature. It was observed that zinc phosphide can be successfully synthesized using this simple chemical route in both nano-wire and thin film forms depending on the exposure mechanism of the phosphorous precursor with the zinc-deposited substrates.

Current Status

Recently Kimball et al have studied the effect of magnesium doping on Zn_3P_2 thin films [98,99]. Our group has recently demonstrated that zinc phosphide can be successfully synthesized in both nano-wire and thin film forms [96,97]. Efforts are ongoing to improve the photovoltaic properties using various hetero-junction partner options.

IRON DISULFIDE (FES_2)

Introduction

Pyrite (FeS_2) is a non-toxic material and is abundant on earth, and has a potential to be an inexpensive and sustainable alternative for achieving low-cost and high-efficiency solar cells due both to its environmental compatibility and its optimal optical properties for efficient energy conversion. Its energy band gap of 0.95 eV and optical absorption coefficient of the order of 10^5 cm^{-1} are in the optimal range for efficient conversion of sunlight into electricity [12,100]. A saturation current of 40 to 50 mA/cm^2 for a thickness between 0.2 μm to 1 μm is possible for pyrite [101]. In addition, a carrier mobility of 200 cm^2/Vs and sufficiently high minority carrier lifetimes makes pyrite highly favorable for photovoltaic application [102]. It has been calculated that theoretical conversion efficiency close to 30% is possible because its band gap is quite close to silicon.

Iron pyrite, like other metal chalcogenides, received considerable attention after the photoelectrochemical solar cells using FeS_2 single crystal electrodes in contact with iodide/tri-iodide (I^-/I_3^-) redox electrolyte exhibited a quantum efficiency close to 100% across a n-FeS_2/(I^-/I_3^-) interface [103-105]. However, a solar conversion efficiency of this photochemical cell was only 2.8% when illuminated under AM 1.5. The open circuit voltage (V_{oc}) was 187 mV. This value of V_{oc} is lower than the theoretical value of about 500 mV [106]. The low photo-potential is attributed to the strong pinning of the Fermi level caused by surface states [105] as well as bulk defects caused by the sulphur deficiency [107].

It should be noted that numerous iron sulphides exist in nature with unique properties depending on different iron (Fe) and sulphur (S) stoichiometric ratios and different crystal structures. Thus creating a phase pure iron pyrite is a challenge. Orthorhombic marcasite FeS_2 and hexagonal troilite FeS are both common iron sulfur phases, but because they have much smaller band gaps (0.34 eV for marcasite and

0.04 eV for troilite), even trace amounts would significantly diminish the photovoltaic properties of the pyrite [108]. So, examining the preparation of pyrite by using various techniques is important.

Synthesis Techniques

Some of the experimental techniques used to grow pyrite are summarized below:

Chemical Vapor Transport (CVT)

This technique involves high crystal growth with halogens, such as Br_2 and Cl_2, and polycrystalline FeS_2 sealed in an ampoule [109]. This process produced cubic crystals with (111) and (100) faces up to 5 mm edge length. However, the growth process was too long (10 days) and was performed at high temperature (~1000°C). At elevated temperatures, segregation of sulphur and iron species is unavoidable.

Metal Organic Chemical Vapor Depositon (MOCVD)

Iron source, such as iron penta-carbonyl [Fe $(CO)_5$], and sulphur sources such as sulphur and H_2S gas, are used in this technique [102]. The advantage of this technique is that the growth temperature is low (150°C) and the growth process is fast.

Spray Pyrolysis

Chemical spray pyrolysis is used to grow FeS_2 using thiourea and $FeCl_2$ [110]. Chlorine contamination, slow growth, non-uniformity, reduced repeatability are some issues associated with this technique.

Chemical Methods

A recent synthesis approach to grow a single phase FeS_2 nanoparticle is carried out by different groups using hydrothermal process [108,111]. Wadia and his group show single phase growth of FeS_2nanoparticles. On the other hand, colloidal FeS_2 nano-crystal ink by using a hot-injection method has been synthesized by many groups [112-116]. These processes are encouraging in that these are low temperature processes and variation in particle sizes allows the manipulation of band gap, which will help to absorb a broader spectrum of sunlight. To translate these nanocrystals into a form of a compact thin film layer is a challenge [113,116,117].

Sulfurization of Iron Oxides

Sulphurization of iron oxides (Fe_3O_4 or Fe_2O_3) is predicted by Gibbs free energy phase diagram to give FeS_2 films [118]. Smestad et. al, sulphurized Fe_3O_4 or Fe_2O_3 using gaseous sulphur at 350°C. However, they didn't observe a good photovoltaic behavior from the photochemical cells made by using the FeS_2 films grown this way. One of the main reasons attributed to the poor photocurrent and voltage is the presence of micro-pinholes, which led to a short circuit in the photogenerated current and effectively shunted the pyrite cell. Recently, defect-free pyrite thin films were synthesized from iron oxide by using non-toxic and a more controllable organic precursor such as Di-tert butyl disulfide (TBDS) [119].

Sulfurization of Iron

It is also claimed by another group that iron oxide route is not necessary to grow pyrite [12]. They used iron film and sulphurized it under nitrogen flux. But their films had a presence of an amorphous phase, which caused an indirect transion at 1.31 eV incompatible with the strong absorption of FeS_2

Issues to Be Addressed For an Optimal Fes_2 Cell

Low open circuit (V_{oc}) in pyrite solar cell is attributed to sulphur deficiency among other reasons [120]. A correlation between S deficiency and the transport parameters of pyrite was found, indicating that high quality intrinsic material can be prepared when the S/Fe ratio is close to the stoichiometric value of 2 [121]. In addition to the intrinsic material property, a device design would also affect V_{oc}.

Influence of Cell Design on V_{oc}

The electrical and optical properties of FeS_2 are promising for an efficient photovoltaic action, but the conversion efficiency of the FeS_2 cell is not impressive: 1% with Schottky type solar cells [101] and 2.8% with photochemical cells. Cells fabricated with Schottky contacts exhibited short circuit currents (J_{sc}) below 10 mA/cm^2 and open circuit voltages (V_{oc}) below 20 mV. However, quantum efficiencies of 90% and above and high photo-current density observed in the latter type of the cell indicate the potential of an efficient pyrite based solar cell. The reason for the low efficiency was due to low open circuit voltage of about 200 mV. The interfacial chemistry of the p-n junction plays a role in dictating the open circuit voltage. A solar cell structure with a wide band gap window layer, such as ZnS, may improve the open circuit voltage.

Stoichiometry of Fes_2

The ratio of Fe and S is extremely important to obtain a defect-free FeS_2 phase. It is necessary that the homogeneity range should be within 1%. It was shown in pyrite grown by MOCVD that whenever a stoichiometry is such that the composition of S is less than 2, a phase mixture of pyrite and pyrrhotite (FeS) existed. The band gap of pyrrhotite phase is 0.04 eV, which greatly diminishes the photovoltaic action. Thus reaffirmation of the pyrite stoichiometry is important.

Influence of Doping on V_{oc}

A recent toil on simulation of p-n type FeS_2 homo-junction diffusion limited solar cells showed that efficiencies around 20% are possible [101] by assuming that at low doping density Shockley–Read–Hall (SRH) recombination [122,123] limits the carrier life time, or in other words, the cell efficiency. Altermatt et. al [101] simulated a case where the carrier life time (t) limits only V_{oc}, e.g. a cell with d = 1 μm and t = 100 ns. The optimally obtainable photo-current in pyrite under AM 1.5G illumination is about 50 mA/cm^2. The achievable efficiency level then was 18.5% (13.5%) in a highly (lowly) doped cell. There is trade-off between the carrier life time and open circuit voltage. For a high density of carriers, the voltage across the junction increases, which may create a higher V_{oc} provided the recombination of the generated carriers is prevented. This is possible with the low carrier density (thus low dopant induced defects) case where the carrier life time is longer. Thus a subtle balance between V_{oc} and excess carrier life time should be sought by carefully choosing the right dopant density. It was observed in Cobalt doped FeS_2 that by manipulating the carrier concentration, the grain barrier height can be changed, which in turn will change the photo-voltage [124].

Fermi Level Pinning At the Surface

Low photo-voltage in a pyrite solar cell is dependent upon its surface structure. Ultraviolet photoemission spectroscopy (UPS) measurements carried out in Cobalt doped FeS_2 showed that the Fermi level at the surface is pinned [125], implying high density of surface. Thus controlling the dopant density with a high level of precision is critical. In a situation where surface defects are unavoidable, application of a resistive thin layer, such as n-type i-ZnO, could be helpful.

Recent Progress in Fes_2 Solar Cell Device

Recently Ganta et. al fabricated solid state based FeS_2 thin film solar cell using CdS as an n-type layer [116]. The super-strate type cell structure was glass/FTO/CdS/FeS_2/Ag/Cr. FeS_2 layer was formed by drop casting the FeS_2 nanoparticle ink directly onto the CdS layer at room temperature. The efficiency for CdS/FeS_2 cell was 0.03%. The low efficiency was due to the low short-circuit current. The open-circuit voltage was high, which is large compared to a previously reported value of 200 mV [106].

Although the circuit current was low at 0.01 mA/cm^2, an open-circuit voltage of 565 mV was observed. Residual organic components present in the nanoparticle-film could be a reason for low short-circuit current. A compact film with the organic residues removed can be expected to improve the short-circuit current and the cell efficiency.

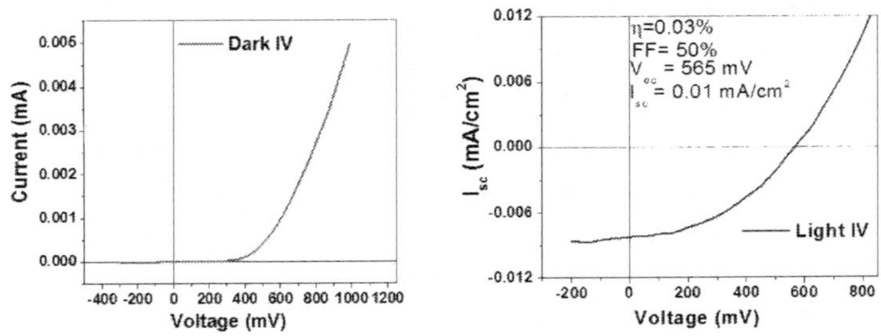

Figure 5: Dark and light I-Vcurves of the FeS_2 super-strate solar cell.

OTHER EARTH-ABUNDANT CANDIDATES

In a recent study, a number of materials were screened based on their abundance and non-toxicity, and semiconducting compounds composed of these identified materials were listed [126]. The authors found that even if binary compounds are focused, there are a good number of alternatives. Once semiconducting properties such as ideal band-gap and minority carrier diffusion length are considered, the next step is the processing capability. Obtaining the right phase without any defects is crucial. The next step is to identify ideal hetero-juction partner with right band-alignment as well as chemical and mechanical compatibility. Several other earth-abundant candidates are being researched these days.

Tin Sulfide (SnS) is metal chalcogenide semiconductor material from the IV-VI group. It has a near optimum direct bandgap of ~ 1.3 eV [127] and also a good absorption co-efficient (10^5 cm^{-1}) [128]. Elements such as tin and sulfur are abundantly available. Tin sulfide has been synthesized using a variety of techniques such as sputtering [127], evaporation [129], electrodeposition [130] as well as chemical routes [131]. The highest reported efficiency to date is 1.3% [132].

Cu_2FeSnS_4 (CFTS) with Stannite structure can also be a viable alternative and there are some recent reports of growing CFTS nanocrystals [133,134]. The bandgap of CFTS was found to be ~ 1.28 eV with p-type conductivity [134].

In a recent study, Fernandes et al identified ternary sulfides Cu_2SnS_3 and Cu_3SnS_4 as good alternatives [135]. The estimated bandgap values are between 1 to 1.6 eV and absorption coefficient 10^4 cm^{-1}.

Cu_2S is another interesting material that was extensively researched in the eighties. However, performance degradation due to copper diffusion led this technology to be abandoned. Cu_2S has a bandgap of 1.2 eV [136]. The highest reported efficiency was 9.1% [137]. Recently this material has again received attention and innovative approaches, such as nanocrystals, are being implemented to overcome the issues encountered in the past [138].

CONCLUSIONS

One of the important p-type absorber CZTSe(S) has reached a laboratory level efficiency of more than 10% and has matured to a stage where it can be seriously considered for commercialization. Complex phase diagram and narrow stoichilometry window for CZTS remains a challenge; in addition, interface properties with suitable hetero-junction partners need to be understood better. However, despite these challenges, CZTS is the most promising material synthesized using earth-abundant constituents. Other materials discussed in detail in this review are Zn_3P_2 and FeS_2. These two materials also show considerable promise, despite their current low lab-level efficiency values. Apart from these, there are several other promosing materials synthesized using earth-abundant constutuents, such as SnS, CFTS and Cu_2SnS_3, and these potentially can be used in solar cells due to their photovoltaic properties.

REFERENCES

1. J. J. J. Loferski, 1956 Theoretical Considerations Governing the Choice of the Optimum Semiconductor for Photovoltaic Solar Energy Conversion J Appl Phys 27 777
2. W.W. Shockley, 1961 Detailed Balance Limit of Efficiency of p-n Junction Solar Cells J Appl Phys 32 510

3. M. A. Green, K. Emery, Y. Hishikawa, W. Warta, E. D. Dunlop, 2011 Solar cell efficiency tables (Version 38) Prog Photovoltaics Res Appl 19 565 72
4. S.S. Guha, 1986 Enhancement of open circuit voltage in high efficiency amorphous silicon alloy solar cells Appl Phys Lett 49 218
5. I. Gordon, L. Carnel, D. Van Gestel, G. Beaucarne, J. Poortmans, 2007 Efficient thin-film polycrystalline-silicon solar cells based on aluminum- induced crystallization and thermal CVD Prog Photovoltaics Res Appl 15 575 86
6. P. Jackson, D. Hariskos, E. Lotter, S. Paetel, R. Wuerz, R. Menner, et al. 2011 New world record efficiency for Cu(In,Ga)Se2 thin-film solar cells beyond 20% Prog Photovoltaics Res Appl 19 894 7
7. X. X. Wu, 2004 High-efficiency polycrystalline CdTe thin-film solar cells Solar Energy 77 803 14
8. H. Katagiri, K. Jimbo, W. S. Maw, K. Oishi, M. Yamazaki, H. Araki, et al. 2009 Development of CZTS-based thin film solar cells Thin Solid Films 517 2455 60
9. C. C. Wadia, 2009 Materials Availability Expands the Opportunity for Large-Scale Photovoltaics Deployment Environ Sci Technol 43 2072 7
10. F. Liu, Y. Li, K. Zhang, B. Wang, C. Yan, Y. Lai, et al. 2010 In situ growth of Cu2ZnSnS4 thin films by reactive magnetron co-sputtering Solar Energy Mater Solar Cells 94 2431 4
11. P. P. S. Nayar, 1981 Zinc phosphide-zinc oxide heterojunction solar cells Appl Phys Lett 39 105
12. B. Rezig, H. Dahman, M. Kenzari, 1992 Iron pyrite FeS2 for flexible solar cells Renewable Energy 2 125 8
13. V. S. Babu, P. R. Vaya, J. Sobhanadri, 1988 Optical absorption and photoconductivity studies on Zn3P2 thin films grown by hot wall deposition Solar Energy Materials 18 65 73
14. H. H. Katagiri, 2001 Development of thin film solar cell based on Cu2ZnSnS4 thin films Solar Energy Mater Solar Cells 65 141 8
15. A. M. Hermann, A. Madan, M. W. Wanlass, V. Badri, R. Ahrenkiel, S. Morrison, et al. 2004 MOCVD growth and properties of Zn3

P2 and Cd3P2 films for thermal photovoltaic applications Solar Energy Mater Solar Cells 82 241 52

16. M. M. Morsli, 1995 Electrical Properties of a Synthetic Pyrite FeS Non Stoichiometric Crystal Journal de physique.I : JP 5 699 705
17. S.S. Sudhakar, 2008 Influence of cooling rate on the liquid-phase epitaxial growth of Zn3 P2 J Cryst Growth 310 2707 11
18. N. G. Dhere, 2007 Toward GW/year of CIGS production within the next decade Solar Energy Materials and Solar Cells 91 1376 82
19. N. G. Dhere, 2011 Scale-up issues of CIGS thin film PV modules Solar Energy Mater Solar Cells 95 277 80
20. M. M. A. Green, 2006 Consolidation of thin-film photovoltaic technology: the coming decade of opportunity Progress in photovoltaics 14 383 92
21. B. A. Anderson, C. Azar, J. Holmberg, S. Karlsson, 1998 Material constraints for thin-film solar cells Energy 23 407 11
22. H. Katagiri, 2005 Cu2ZnSnS4 thin film solar cells Thin Solid Films 480 426
23. A. A. Fischereder, 2010 Investigation of Cu ZnSnS Formation from Metal Salts and Thioacetamide Chemistry of materials 22 3399 406
24. H. Katagiri, N. Sasaguchi, S. Hando, S. Hoshino, J. Ohashi, T. Yokota, 1997 Preparation and evaluation of Cu2ZnSnS4 thin films by sulfurization of E---B evaporated precursors Solar Energy Mater Solar Cells 49 407 14
25. H. Araki, 2008 Preparation of Cu2ZnSnS4 thin films by sulfurization of stacked metallic layers Thin Solid Films 517 1457
26. J. Seol, S. Lee, J. Lee, H. Nam, K. Kim, 2003 Electrical and optical properties of Cu2ZnSnS4 thin films prepared by rf magnetron sputtering process Solar Energy Mater Solar Cells 75 155 62
27. T. Tanaka, D. Kawasaki, M. Nishio, Q. Guo, H. Ogawa, 2006 Fabrication of Cu2ZnSnS4 thin films by co-evaporation physica status solidi (c) 3 2844 7
28. T. Tanaka, T. Nagatomo, D. Kawasaki, M. Nishio, Q. Guo, A. Wakahara, et al. 2005 Preparation of Cu2ZnSnS4 thin films by hybrid sputtering Journal of Physics and Chemistry of Solids 66 1978 81

29. H. Araki, Y. Kubo, A. Mikaduki, K. Jimbo, W. S. Maw, H. Katagiri, et al. 2009 Preparation of Cu2ZnSnS4 thin films by sulfurizing electroplated precursors Solar Energy Mater Solar Cells 93 996 9
30. K. Moriya, J. Watabe, K. Tanaka, H. Uchiki, 2006 Characterization of Cu2ZnSnS4 thin films prepared by photo-chemical deposition physica status solidi (c) 3 2848 52
31. K. Tanaka, M. Oonuki, N. Moritake, H. Uchiki, 2009 Cu2ZnSnS4 thin film solar cells prepared by non-vacuum processing Solar Energy Mater Solar Cells 93 583 7
32. M. M. Y. Yeh, 2009 Influences of synthesizing temperatures on the properties of Cu2ZnSnS4 prepared by sol-gel spin-coated deposition J Sol Gel Sci Technol 52 65 8
33. N. Nakayama, K. Ito, 1996 Sprayed films of stannite Cu2ZnSnS4 Appl Surf Sci 92 171 5
34. N. Kamoun, H. Bouzouita, B. Rezig, 2007 Fabrication and characterization of Cu2ZnSnS4 thin films deposited by spray pyrolysis technique Thin Solid Films 515 5949 52
35. S. C. Riha, B. A. Parkinson, A. L. Prieto, 2009 Solution-Based Synthesis and Characterization of Cu2ZnSnS4 Nanocrystals J Am Chem Soc 131 12054 5
36. T. K. Todorov, K. B. Reuter, D. B. Mitzi, 2010 High-Efficiency Solar Cell with Earth-Abundant Liquid-Processed Absorber Adv Mater 22 E 156 9
37. T.N. K. Ito, 1988 Electrical and optical properties of stannite-type quaternary semiconductor thin films Japanese journal of Applied Physics
38. D. A. R. Barkhouse, O. Gunawan, T. Gokmen, T. Todorov, D. Mitzi, 2012 Device characteristics of a 10.1% hydrazine-processed Cu2ZnSn(Se,S)4 solar cell Characteristics of a 10.1% efficient kesterite solar cell Progress in photovoltaics 20 6 11
39. B. Shin, O. Gunawan, Y. Zhu, N. A. Bojarczuk, S. J. Chey, S. Guha, 2011 Thin film solar cell with 8.4% power conversion efficiency using an earth-abundant Cu2ZnSnS4 absorber Prog Photovoltaics Res Appl:n/a,n/a
40. B. B. J. Stanbery, 2002 Copper Indium Selenides and Related Materials for Photovoltaic Devices Critical reviews in solid state and materials sciences 27 73 117

41. S. S. Siebentritt, 2012 Kesterites-a challenging material for solar cells Kesterites-a challenging material for solar cells Progress in photovoltaics:n,a-n/a
42. I. D. Olekseyuk, 2004 Phase equilibria in the Cu2S-ZnS-SnS2 system J Alloys Compounds 368 135 43
43. H. Katagiri, 2008 Enhanced conversion efficiencies of Cu2ZnSnSP4based thin film solar cells by using preferential etching technique Applied physics express 1
44. H. Katagiri, K. Jimbo, K. Moriya, K. Tsuchida, 2003 Solar cell without environmental pollution by using CZTS thin film Photovoltaic Energy Conversion, Proceedings of 3rd World Conference on 3 2874 2879
45. H. Katagiri, K. Jimbo, 2011 Development of rare metal-free CZTS-based thin film solar cells Photovoltaic Specialists Conference (PVSC), 37th IEEE 003516 21
46. K. Wang, O. Gunawan, T. Todorov, B. Shin, S. J. Chey, N. A. Bojarczuk, et al. 2010 Thermally evaporated Cu2ZnSnS4 solar cells Appl Phys Lett 97 143508
47. B. Shin, Y. Zhu, N. A. Bojarczuk, S. J. Chey, S. Guha, 2012 High efficiency Cu2ZnSnSe4 solar cells with a TiN diffusion barrier on the molybdenum bottom contact 38th IEEE Photovoltaic Specialists Conference (PVSC)
48. I. Repins, C. Beall, N. Vora, C. De Hart, D. Kuciauskas, P. Dippo, et al. 2012 Co-evaporated Cu2ZnSnSe4 films and devices Solar Energy Mater Solar Cells 101 154 9
49. V. Chawla, B. Clemens, 2012 Effect of Composition on High Efficiency CZTSSe Devices Fabricated Using Co-sputtering of Compound Targets 38th IEEE Photovoltaic Specialists Conference (PVSC)
50. H. Hiroi, N. Sakai, S. Muraoka, T. Katou, H. Sugimoto, 2012 Development of High Efficiency Cu2ZnSnS4 Submodule with Cd-Free Buffer Layer 38th IEEE Photovoltaic Specialists Conference (PVSC)
51. H. Sugimoto, H. Hiroi, N. Sakai, S. Muraoka, T. Katou, 2012 Over 8% Efficiency Cu2ZnSnS4 Submodules with Ultra-Thin Absorber 38th IEEE Photovoltaic Specialists Conference (PVSC)

52. P. M. P. Salomé, J. Malaquias, P. Fernandes, M. Ferreira, J. Leitao, A. Cunha, et al. 2011 The influence of hydrogen in the incorporation of Zn during the growth of Cu2ZnSnS4 thin films Solar Energy Mater Solar Cells 95 12 3482 9
53. S. W. Shin, S. Pawar, C. Park, J. Yun, J. Moon, J. Kim, et al. 2011 Studies on Cu2ZnSnS4 (CZTS) absorber layer using different stacking orders in precursor thin films Solar Energy Mater Solar Cells 95 12 3202 6
54. R. B. V. Chalapathy, G. S. Jung, B. T. Ahn, 2011 Fabrication of Cu2ZnSnS4 films by sulfurization of Cu/ZnSn/Cu precursor layers in sulfur atmosphere for solar cells Solar Energy Mater Solar Cells 95 12 3216 21
55. S. Aksu, S. Pethe, A. Kleiman-Shwarsctein, S. Kundu, M. Pinarbasi, 2012 Recent Advances in Electroplating Based CIGS Solar Cell Fabrication 38th IEEE Photovoltaic Specialists Conference (PVSC)
56. J. J. Scragg, P. J. Dale, L. M. Peter, G. Zoppi, I. Forbes, 2008 New routes to sustainable photovoltaics: evaluation of Cu2ZnSnS4 as an alternative absorber material physica status solidi (b) 245 1772 8
57. J. J. Scragg, D. M. Berg, P. J. Dale, 2010 A 3.2% efficient Kesterite device from electrodeposited stacked elemental layers J Electroanal Chem 646 52 9
58. A. Ennaoui, 2009 Cu2ZnSnS4 thin film solar cells from electroplated precursors: Novel low-cost perspective Thin Solid Films 517 2511
59. H. Araki, Y. Kubo, K. Jimbo, W. Maw, H. Katagiri, M. Yamazaki, et al. 2009 Preparation of Cu2ZnSnS4 thin films by sulfurization of co-electroplated Cu-Zn-Sn precursors physica status solidi (c) 6 1266 8
60. J. Li, T. Wei, M. Liu, W. Jiang, G. C. Zhu, 2012 The Cu2ZnSnSe4 thin films solar cells synthesized by electrodeposition route Appl Surf Sci 258 6261 5
61. S. S. Mali, P. S. Shinde, C. A. Betty, P. N. Bhosale, Y. W. Oh, P. S. Patil, 2012 Synthesis and characterization of Cu2ZnSnS4 thin films by SILAR method Journal of Physics and Chemistry of Solids 73 735 40

62. J. J. B. Mooney, 1982 Spray Pyrolysis Processing Annual review of materials science 12 81 101
63. T. Prabhakar, N. Jampana, 2011 Effect of sodium diffusion on the structural and electrical properties of Cu2ZnSnS4 thin films Solar Energy Mater Solar Cells 95 1001 4
64. V. G. Rajeshmon, C. S. Kartha, K. P. Vijayakumar, C. Sanjeeviraja, T. Abe, Y. Kashiwaba, 2011 Role of precursor solution in controlling the opto-electronic properties of spray pyrolysed Cu2ZnSnS4 thin films Solar Energy 85 249 55
65. K. Tanaka, N. Moritake, H. Uchiki, 2007 Preparation of thin films by sulfurizing sol-gel deposited precursors Solar Energy Mater Solar Cells 91 1199 201
66. Z. Zhou, Y. Wang, D. Xu, Y. Zhang, 2010 Fabrication of Cu2ZnSnS4 screen printed layers for solar cells Solar Energy Mater Solar Cells 94 2042 5
67. C. Chory, F. Zutz, F. Witt, H. Borchert, J. Parisi, 2010 Synthesis and characterization of Cu2ZnSnS4 physica status solidi (c) 7 1486 8
68. Q. Guo, G. M. Ford, W. Yang, C. J. Hages, H. W. Hillhouse, R. Agrawal, 2012 Enhancing the performance of CZTSSe solar cells with Ge alloying Solar Energy Mater Solar Cells 105 132 6
69. Q. Guo, Y. Cao, J. Caspar, W. Farneth, A. Ionkin, L. Johnson, et al. 2012 A Simple Solution-based Route to High-Efficiency CZTSSe Thin-film Solar Cells 38th IEEE Photovoltaic Specialists Conference (PVSC)
70. B. Kumar, P. S. Vasekar, S. Pethe, N. G. Dhere, T. Galymzhan, 2009 ZnxCd1-xS as a heterojunction partner for CuIn1-xGaxS2 thin film solar cells Thin Solid Films 517 2295 9
71. G. Shen, P. Chen, Y. Bando, D. Golberg, C. Zhou, 2008 Single-Crystalline and Twinned Zn3P2 Nanowires: Synthesis, Characterization, and Electronic Properties The Journal of Physical Chemistry C 112 16405 10
72. K. Kakishita, K. Aihara, T. Suda, 1994 Zn3P2 photovoltaic film growth for Zn3P2/ZnSe solar cell Solar Energy Mater Solar Cells 35 333 40

73. J. Misiewicz, L. Bryja, K. Jezierski, J. Szatkowski, N. Mirowska, Z. Gumienny, et al. 1994 Zn3P2 new material for optoelectronic devices Microelectron J 25 23 28
74. A. Lousa, E. Bertrán, M. Varela, J. L. Morenza, 1985 Deposition of Zn3P2 thin films by coevaporation Solar Energy Materials 12 51 6
75. R. Sathyamoorthy, C. Sharmila, K. Natarajan, S. Velumani, Influence of annealing on structural and optical properties of Zn3P2 thin films Mater Charact 58 745 9
76. K. R. Murali, D. R. Rao, 1981 Optical band gap of [alpha]- Zn3P2 Thin Solid Films 86 283 6
77. J. Long, 1983 The Growth of Zn3P2 by Metalorganic Chemical Vapor Deposition J Electrochem Soc 130 725 8
78. M. M. Bhushan, A. Catalano, 1981 Polycrystalline Zn3P2 Schottky barrier solar cells Appl Phys Lett 38 39 41
79. K. Kakishita, 1994 Zinc phosphide epitaxial growth by photo-MOCVD Appl Surf Sci 79 281
80. A. Catalano, 1980 Defect dominated conductivity in Zn3P2 The Journal of physics and chemistry of solids 41 635
81. M. S. Casey, A. L. Fahrenbruch, R. H. Bube, 1987 Properties of zinc-phosphide junctions and interfaces Journal of Applied Physics 61 2941 6
82. M.Bhushan,J.A.Turner,B.A.Parkinson,1986Photoelectrochemical Investigation of Zn3P2 J Electrochem Soc 133 536 9
83. M. M. Bhushan, 1982 Schottky solar cells on thin polycrystalline Zn3P2 films Appl Phys Lett 40 51
84. N. C. Wyeth, A. Catalano, 1979 Spectral response measurements of minority-carrier diffusion length in Zn3P2 Journal of Applied Physics 50 1403 7
85. M. M. Ginting, J. Leslie, 1989 Preparation and electrical properties of heterojunctions of ZnO on Zn3P2 and CdTe Can J Phys 67 448 55
86. T. Suda, M. Suzuki, S. Kurita, 1983 Polycrystalline Zn3P2 Indium-Tin Oxide Solar Cells Jpn J Appl Phys 22 L656 8
87. A. A. Catalano, 1980 Evidence of p/n homojunction formation in Zn3P2 Appl Phys Lett 37 567

88. M. M. Bhushan, 1982 Mg diffused zinc phosphide n/p junctions J Appl Phys 53 514
89. J. Misiewicz, J. Szatkowski, N. Mirowska, K. Jezierski, Z. Gumienny, 1986 Solar energy conversion in metal- Zn3P2 contacts Acta Physica Polonica A69 1131 5
90. A. A. Catalano, 1980 The growth of large Zn3P2 crystals by vapor transport J Cryst Growth 49 681 6
91. Faa-Ching Wang, R. H. Bube, R. S. Feigelson, R. K. Route, 1981 Single crystal growth of Zn3P2 J Cryst Growth 55 268 72
92. J. J. L. Deiss, 1988 Amorphous thin films of Zn3P2 Phys Scripta 37 587 92
93. K. R. Murali, 1987 Properties of Zn3P2 films grown by close space evaporation. Materials science & engineering A, Structural materials : properties, microstructure and processing 92 193
94. M. Soliman, A. B. Kashyout, M. Osman, M. El-Gamal, 2005 Electrochemical deposition of Zn3P2 thin film semiconductors on tin oxide substrates Renewable Energy 30 1819 29
95. Y. Y. Nose, 2012 Electrochemical Deposition of Zn3P2 Thin Film Semiconductors Based on Potential-pH Diagram of the Zn-P-H2O System J Electrochem Soc 159 D181
96. P. S. Vasekar, S. P. Adusumilli, D. Vanhart, T. Dhakal, S. Desu, 2012 Development of zinc phosphide as a p-type absorber Photovoltaic Specialists Conference (PVSC), 38th IEEE
97. S. P. Adusumilli, P. S. Vasekar, D. Vanhart, T. Dhakal, C. R. Westgate, S. Desu, 2011 Development of zinc phosphide as an absorber using chemical reflux technique MRS conference proceedings, Spring
98. G. M. Kimball, A. Müller, N. S. Lewis, H. A. Atwater, 2009 Photoluminescence-based measurements of the energy gap and diffusion length of Zn3P2 Appl Phys Lett 95 112103
99. G. M. Kimball, N. S. Lewis, H. A. Atwater, 2010 Mg doping and alloying in Zn3P2 heterojunction solar cells Photovoltaic Specialists Conference (PVSC), 35th IEEE 001039-43
100. J. P. Wilcoxon, P. P. Newcomer, G. A. Samara, 1996 Strong quantum confinement effects in semiconductors: FeS2 nanoclusters Solid State Commun 98 581 5

101. P. P. Altermatt, T. Kiesewetter, K. Ellmer, H. Tributsch, 2002 Specifying targets of future research in photovoltaic devices containing pyrite (FeS2) by numerical modelling Solar Energy Mater Solar Cells 71 181 95
102. G. G. Chatzitheodorou, 1986 Thin photoactive FeS2 (pyrite) films Mater Res Bull 21 1481 7
103. A. A. Ennaoui, 1985 Photoactive Synthetic Polycrystalline Pyrite (FeS2) J Electrochem Soc 132 1579
104. A. Ennaoui, 1993 Iron disulfide for solar energy conversion Solar Energy Mater Solar Cells 29 289
105. A. Ennaoui, S. Fiechter, G. Smestad, H. Tributsch, 1990 Preparation of iron disulphide and its use for solar energy conversion Proceedings of 1st World Renewable Energy Congress 458 64
106. H. H. Dahman, 1996 Iron pyrite films prepared by sulfur vapor transport Thin Solid Films 280 56 60
107. S. Fiechter, 1992 The microstructure and stoichiometry of pyrite J Mater Res 7
108. C. C. Wadia, 2009 Surfactant-Assisted Hydrothermal Synthesis of Single phase Pyrite FeS Nanocrystals Chemistry of materials 21 2568 70
109. M. Kunst, H. Tributsch, 1984 Surface microwave conductivity spectroscopy. A new experimental tool Chemical Physics Letters 105 123 6
110. G. Smestad, A. Da Silva, H. Tributsch, S. Fiechter, M. Kunst, N. Meziani, et al. 1989 Formation of semiconducting iron pyrite by spray pyrolysis Solar Energy Materials 18 299 313
111. R. Wu, Y. F. Zheng, X. G. Zhang, Y. F. Sun, J. B. Xu, J. K. Jian, 2004 Hydrothermal synthesis and crystal structure of pyrite J Cryst Growth 266 523 7
112. J. J. Joo, 2003 Generalized and Facile Synthesis of Semiconducting Metal Sulfide Nanocrystals J Am Chem Soc 125 11100 5
113. Y. Y. Y. Lin, 2009 Extended red light harvesting in a poly(P3-hexylthiophene)/iron disulfide nanocrystal hybrid solar cell Nanotechnology 20 405207
114. Y. Y. Bi, 2011 Air Stable, Photosensitive, Phase Pure Iron Pyrite Nanocrystal Thin Films for Photovoltaic Application Nano letters 11 4953 7

115. J. J. Puthussery, 2010 Colloidal Iron Pyrite (FeS) Nanocrystal Inks for Thin-Film Photovoltaics J Am Chem Soc 101222110759049

116. L. Ganta, T. Dhakal, S. Rajendran, C. R. Westgate, 2012 Synthesis of FeS2 nanocrystals for ink-based solar cells MRS Online Proceedings Library 1447

117. T. Dhakal, L. Ganta, D. Vanhart, C. R. Westgate, 2012 Annealing of FeS2 Nano-crystal Thin Film 38th IEEE Photovoltaic Specialists Conference (PVSC)

118. G. Smestad, A. Ennaoui, S. Fiechter, H. Tributsch, W. K. Hofmann, M. Birkholz, et al. 1990 Photoactive thin film semiconducting iron pyrite prepared by sulfurization of iron oxides Solar Energy Materials 20 149 65

119. S. P. Adusumilli, T. Dhakal, 2012 Synthesis of Iron Pyrite Film through Low Temperature Atmospheric Pressure Chemical Vapor Deposition MRS conference proceedings, Spring

120. N. Alonso-Vante, G. Chatzitheodorou, S. Fiechter, N. Mgoduka, I. Poulios, H. Tributsch, 1988 Interfacial behavior of hydrogen-treated sulphur deficient pyrite (FeS2−x) Solar Energy Materials 18 9 21

121. R. R. Schieck, 1990 Electrical properties of natural and synthetic pyrite (FeS2) crystals J Mater Res 5 1567 72

122. R. R. N. Hall, 1952 Electron-Hole Recombination in Germanium Physical review 87 387

123. W. W. Shockley, 1952 Statistics of the Recombinations of Holes and Electrons Physical review 87 835 42

124. J. Oertel, K. Ellmer, W. Bohne, J. Röhrich, H. Tributsch, 1999 Growth of n-type polycrystalline pyrite (FeS2) films by metalorganic chemical vapour deposition and their electrical characterization J Cryst Growth 198-199 Part 2 1205 10

125. K. Ellmer, H. Tributsch, 2000 Iron disulphide (pyrite) as photovoltaic material: problems and opportunities Proceedings of the 12th Workshop on Quantum Solar Energy Conversion

126. F. Alharbi, J. Bass, A. Salhi, A. Alyamani, H. Kim, R. Miller, 2011 Abundant non-toxic materials for thin film solar cells: Alternative to conventional materials Renewable Energy 36 10 2753 8

127. K. T. Ramakrishna Reddy, P. Purandhara Reddy, P. K. Datta, R. W. Miles, 2002 Formation of polycrystalline SnS layers by a two-step process Thin Solid Films 403 404 116 9
128. M. Devika, N. Reddy, K. Gunasekhar, 2011 Structural, electrical, and optical properties of as-grown and heat treated ultra-thin SnS films Thin Solid Films 520 1 628 32
129. M. Devika, N. Reddy, K. Ramesh, K. Gunasekhar, E. Gopal, K. T. R. Reddy, 2006 Influence of annealing on physical properties of evaporated SnS films Semiconductor science and technology 21 1125 31
130. M. M. Gunasekaran, 2007 Photovoltaic cells based on pulsed electrochemically deposited SnS and photochemically deposited CdS and Cd1−xZnxS Solar Energy Mater Solar Cells 91 774 8
131. D. Avellaneda, M. T. S. Nair, P. K. Nair, 2009 Photovoltaic structures using chemically deposited tin sulfide thin films Thin Solid Films 517 2500 2
132. K. T. Ramakrishna Reddy, N. Koteswara Reddy, R. W. Miles, 2006 Photovoltaic properties of SnS based solar cells Solar Energy Mater Solar Cells 90 3041 6
133. L. Li, X. Liu, J. Huang, M. Cao, S. Chen, Y. Shen, et al. 2012 Solution-based synthesis and characterization of Cu2FeSnS4 nanocrystals Mater Chem Phys 133 688 91
134. C. Yan, C. Huang, J. Yang, F. Liu, J. Liu, Y. Lai, et al. 2012 Synthesis and characterizations of quaternary Cu2FeSnS4 nanocrystals Chem Commun 48 2603 5
135. P. P. A. Fernandes, 2010 A study of ternary Cu2SnS3 and Cu3SnS4 thin films prepared by sulfurizing stacked metal precursors Journal of physics.D, Applied physics 43 215403
136. G. G. Liu, T. Schulmeyer, J. Brötz, A. Klein, W. Jaegermann, 2003 Interface properties and band alignment of Cu2S/CdS thin film solar cells Thin Solid Films 431 432 477 82
137. J. A. Bragagnolo, A. Barnett, J. Phillips, R. Hall, A. Rothwarf, 1980 The design and fabrication of thin-film CdS/Cu2S cells of 9.15-percent conversion efficiency IEEE Trans Electron Devices 27 645 51

138. Y. Y. Wu, C. Wadia, W. Ma, B. Sadtler, A. P. Alivisatos, 2008 Synthesis and Photovoltaic Application of Copper(I) Sulfide Nanocrystals Nano letters 8 2551 5

Citations

CHAPTER 1

Sadia Ameen, M. Shaheer Akhtar, Minwu Song and Hyung Shik Shin (2013). Metal Oxide Nanomaterials, Conducting Polymers and Their Nanocomposites for Solar Energy, Solar Cells - Research and Application Perspectives, Prof. Arturo Morales-Acevedo (Ed.), ISBN: 978-953-51-1003-3, InTech, DOI: 10.5772/51432.

CHAPTER 2

Bernd Weise,.Impact of K-factor and active current reduction during fault-ride-through of generating units connected via voltage-sourced converters on power system stability, DOI: 10.1049/iet-rpg.2014.0116.

CHAPTER 3

Xiang-Dong Gao, Xiao-Min Li, and Xiao-Yan Gan (2013). Enhancing the Light Harvesting Capacity of the Photoanode Films in Dye-Sensitized Solar Cells, Solar Cells - Research and Application Perspectives, Prof. Arturo Morales-Acevedo (Ed.), ISBN: 978-953-51-1003-3, InTech, DOI: 10.5772/51633.

CHAPTER 4

Tayyar Dzhafarov (2013). Silicon, Solar Cells with Nanoporous Silicon Layer, Solar Cells - Research and Application Perspectives, Prof. Arturo Morales-Acevedo (Ed.), ISBN: 978-953-51-1003-3, InTech, DOI: 10.5772/51593.

CHAPTER 5

Foozieh Sohrabi, Arash Nikniazi, and Hossein Movla (2013), Optimization of Third Generation Nanostructured Silicon- Based Solar Cells, Solar Cells - Research and Application Perspectives, Prof. Arturo Morales-Acevedo (Ed.), ISBN: 978-953-51-1003-3, InTech, DOI: 10.5772/51616.

CHAPTER 6

Eoghan McKenna, Philipp Grünewald, and Murray Thomson, Going with the Wind: Temporal Characteristics of Potential Wind Curtailment in Ireland in 2020 and Opportunities for Demand Response, DOI: 10.1049/iet-rpg.2013.0320.

CHAPTER 7

Parag S. Vasekar and Tara P. Dhakal (2013). Thin Film Solar Cells Using Earth-Abundant Materials, Solar Cells - Research and Application Perspectives, Prof. Arturo Morales-Acevedo (Ed.), ISBN: 978-953-51-1003-3, InTech, DOI: 10.5772/51734.

Index

D

E

F

G

H

I

L

M

T

V

W

X